How to Explain Behavior

How to Explain Behavior

A Critical Review and New Approach

Sam S. Rakover

LEXINGTON BOOKS
Lanham • Boulder • New York • London

Published by Lexington Books
An imprint of The Rowman & Littlefield Publishing Group, Inc.
4501 Forbes Boulevard, Suite 200, Lanham, Maryland 20706
www.rowman.com

Unit A, Whitacre Mews, 26-34 Stannary Street, London SE11 4AB

British Library Cataloguing in Publication Information Available

Library of Congress Cataloging-in-Publication Data

Names: Rakover, Sam S., 1938- author.
Title: How to explain behavior : a critical review and new approach / Sam S. Rakover.
Description: Lanham : Lexington Books, [2017] | Includes bibliographical references and index.
Identifiers: LCCN 2017039648 (print) | LCCN 2017049120 (ebook) | ISBN 9781498536691 (Electronic) | ISBN 9781498536684 (cloth : alk. paper)
Subjects: LCSH: Behavioral assessment. | Explanation.
Classification: LCC BF176.5 (ebook) | LCC BF176.5 .R35 2017 (print) | DDC 150.28/7—dc23
LC record available at https://lccn.loc.gov/2017039648

♾™ The paper used in this publication meets the minimum requirements of American National Standard for Information Sciences—Permanence of Paper for Printed Library Materials, ANSI/NISO Z39.48-1992.

Printed in the United States of America

Contents

Preface

The aims of the book are (a) to offer readers a critical review of the concepts of explanation and understanding as they are applied in science and in the philosophy of science; (b) to examine explanation models current in the natural and social sciences and to compile a list of the basic properties of scientific explanation; (c) to offer psychologists an explanation model based coherently on mechanistic explanations (which are current in the natural sciences) and on mentalistic explanations (which are based on the inner world of the individual: will, belief, consciousness, and awareness); and (d) to offer psychologists a general model that may unite all kinds of explanation given for behavior of the individual.

In the first year of my studies at the Hebrew University in Jerusalem, the psychology department arranged a debate on the motion "Psychology is a science." The speaker against the motion was Professor Yeshayahu Leibowitz, a revered figure not only everywhere on the university campus but also throughout the state of Israel. He was the editor-in-chief of the *Hebrew Encyclopedia*, and after the conquest of the West Bank of the River Jordan in the Six Day War, he uttered a call similar to that made by David Ben-Gurion (former prime minister of Israel) to return the conquered territories because the conquest would cause Israel immense damage and intolerable corruption. (To my great regret, no one paid any attention to these wise personalities.) For the motion were three students: Amos Tversky, Yehoshafat Giveon, and Emanuel Donchin. At the end of the debate all members of the psychology department, lecturers and students, were convinced that Professor Yeshayahu Leibowitz had lost. I, however, felt otherwise. In my opinion, no one had managed to counter his argument that none could feel Professor Yeshayahu Leibowitz's toothache except Leibowitz himself. The response to this argument (given by Emanuel Donchin as best I recall)—that Leibowitz's sensation of pain was of no interest to the psychologist because what was important for the science of psychology was the plain fact observable to all that Professor Leibowitz went to the dentist—seemed to me pretentious, and did not tackle the very problem that the eminent professor had indicated: a very important part of human behavior simply does not accord with the methodology of the natural sciences. In truth, the problem presented by Professor Leibowitz through the toothache case is a well-known and still unresolved problem, the mind/body problem, or the consciousness/brain problem.

This debate left an indelible impression on me, and since that moment I have never ceased thinking about it or reading countless books and articles on the mind/body problem. Chapter 6 of this book makes a methodological attempt not to solve the consciousness/brain problem—but to skirt it! This is an effort to tread a methodological path that circumvents this problem by developing a theoretical framework that will contain both explanations acceptable in science (which I term *mechanistic*) and explanations that treat the individual's inner world, which are anchored to mental states and processes whereby the person's public behavior becomes explicable (explanations I term *mentalistic*). So Emanuel Donchin's response largely reflected the behaviorist view, which disregards the individual's internal, subjective-mental world as an explanatory factor, and concentrates on his behavior that is publicly observable. I propose in chapter 6 to restore the individual's inner world as an essential component in psychological explanations, and show that mentalistic explanations can be part of science's methodological framework.

Another lecture that deeply impressed me was delivered by the late Professor Amos Tversky in the framework of my master's studies. His lecture series was in mathematical psychology, and the subject that influenced me especially was making measurements in science and in psychology. I took great interest in this matter, which resulted in my concluding that the difference between measuring in psychology and measuring in physics was largely responsible for psychology not developing as the natural sciences were developing. Furthermore, I held that these differences between the sciences and psychology arose because in the framework of scientific concepts, it is impossible to understand the conscious experience that enfolds all of us (and, to varying degrees, supreme animals). That is, I suggest that these differences (including the difference in measuring) originate in that to date no one has succeeded in developing a theory explaining how neurophysiological processes in the brain become mental processes, a sense of consciousness, conscious experience (see Rakover, 2012b).

I am bound to add in this context that just a few years ago a lecturer in the psychology department who was giving the basic Introduction to Psychology course asked me to share with the students something about psychology from a philosophical standpoint. I gave them a lecture based on the foregoing thoughts and on the 2012 article (Rakover, 2012b); I thought that the lecture would spark among them a discussion of the question aroused by the psychology department of my youth: Is psychology a science? To my great astonishment, sometime later I was told that that lecturer was annoyed with me, complaining that my talk was out of place because it undermined the students' belief—which he was painstakingly laboring to nurture—that psychology was indeed a science like any science. What a surprise! Only about fifty-five years since that famous debate with Professor Leibowitz, and lo and behold the psychologists have passed from a stance of healthy curiosity—is psychology a science?—to a stance of aggressive defense: yes, we are a science like every science, and woe betide anyone who would tell us otherwise!

In 1990 I published my first book, on the relation between philosophy of science and psychology: *Metapsychology: Missing Links in Behavior, Mind, and Science*. During its writing, this book was read by Professor Leibowitz in the course of an entire year, from cover to cover, chapter by chapter, on which he offered sage and useful comments, and for which in the end he wrote a splendid preface. The book contains inter alia chapters on scientific explanation generally and on giving explanations in psychology particularly. Thereafter, for a quarter of a decade I concentrated more and more on the question of scientific explanation and on the question, "What do psychologists do when they offer explanations in their research?" As a result, I joined Professor Giora Hon of the Department of Philosophy at Haifa University, and for several years we delivered a course for philosophy and psychology students on models of explanation. In addition, we arranged a study day on explanation, with lectures given by the Salmon couple and by Peter Lipton. The lectures of the study day appear in a book that Giora and I edited (Hon & Rakover, 2001).

So here in my hands is the present book, whose composition I completed after long years of thought, and instead of arousing in me a sense of pride and joy, it generates a strange, disconcerted feeling. Imagine, dear reader: On the one hand you can say to yourself that the book presents everything you have ever thought about explanation, more particularly explanation in psychology, and in truth you have set forth in it everything you wanted to say on the subject. This book is not like an article, where you know that your propositions are still partial, that you will still have more to offer on the subject which you have not yet fully exhausted; your head is beginning to reverberate with new ideas or developments of old ones. No: into this book you have poured everything you have had to say up to now, and not even the tiny buds of a new interesting notion are stirring in your head. That should be a wonderful feeling—you have expressed everything you had in your head, everything you wanted to say, everything you thought worth saying. And now, from another direction, you are suddenly overwhelmed by an unnerving sense of uncertainty; perhaps what you have written, what you have expressed after thinking for years, perhaps . . . everything is wrong. Fundamentally wrong. And precisely *because* you have expressed everything you have to say, you are left exposed, because you don't have any more ideas to clutch at like a lifebelt, to save your soul as you gag and drown in the deluge of your thoughts' crassly stupid mistakes. Over the years a riot of ideas has played havoc in your head, and now at the end of the writing, silence has fallen—not necessarily blessed, not necessarily soothing, instead, you tell yourself, a quite frightening silence, a threatening stillness. Perhaps you wanted to go north to drink your fill from the freshwater spring, but no, you have stumbled southward, and the silence around you is nothing but that of parched wilderness. And all you can tell yourself, time after time: look, you tested your thoughts over and over, and some of them, on which the book rests, were published as articles in professional journals, so if there had been mistakes and nonsense, the critics and the editors would have zeroed in on them long ago, so you have nothing to fear, and the ideas expressed in your book are interesting and good, and actually you had no choice and you had to write what you wrote because *c'est la vie.*

Acknowledgments

My heartfelt thanks go all those who read the book in its entirety, or selected chapters of it, and made excellent, useful comments that improved the book's content and reading: Oded Balaban, Benny Beit-Halahmi, Meir Hemmo, Adir Cohen, Motti Rimor, and Aviva Rakover. Special thanks go to Murray Rosovsky, who translated large parts of the book from the original Hebrew, edited the English, and offered some very useful comments, and to Noga Yoselevich, who did the drawings. All have my warmest gratitude. Any mistakes that have survived in the book are nothing other than a kind of behavior that beggars rational explanation.

Introduction

The introduction has two sections. The first, "Basic Ideas: Methodological Credo," treats the ideas and basic premises concerning explanation and understanding, which I acquired over the years starting from the moment I began to take an interest in philosophy of science. Immediately with the onset of my studies at the Hebrew University as a student of psychology, sociology, and statistics, I began auditing courses in philosophy. First came Greek philosophy (including pre-Socratic); then after a while I took classes in the philosophy of science, which delved into the positivist approach, Popper, Kuhn, and so on. The second section outlines the structure of the book, its strengths and weaknesses, and briefly summarizes the content of each chapter.

BASIC IDEAS: METHODOLOGICAL CREDO

It is most reasonable to assume that reading the book will be easier, on the one hand, and deeper and more critical, on the other hand, if I detail in this section the basic principles that guided me in writing the book, that is, if I describe the thought infrastructure on which I built the entire book, most particularly chapters 1, 5, 6, and 7. These views, which appear here (and see details in chapter 4), range from those on the philosophy of explanation in general to those on giving explanations in psychology. Many researchers propose that explanation is anchored to reduction (for reviews, see Jones, 2000; Ney, 2016; van Riel & Van Gulick, 2016). Nevertheless, I will not discuss reduction per se, but mainly examine the attempts to reduce consciousness to neurophysiology (see especially chapters 5 and 6).

I will start the discussion by emphasizing the following two general cornerstone ideas in scientific research methodology (citations are taken from Einstein & Infeld, 1938/1950):

1. *The scientific attempt is to explain and understand reality:* "The purpose of any physical theory is to explain as wide a range of phenomena as possible. It is justified in so far as it does make events understandable" (p. 43). "All the essential ideas in science were born in dramatic conflict between reality and our attempts at understanding" (p. 280).
2. *Scientific knowledge is provisional:* "There are no eternal theories in science. It always happens that some of the facts predicted by a theory are disproved by experiment. Every theory has its period of gradual development and triumph, after which it may experience a rapid decline" (p. 77).

Now I will move forward to discuss six methodological ideas that are more specific to the subject of the book.

1. *To offer an explanation for a studied phenomenon, a suitable scientific procedure has to be used.* The idea underlying giving explanations in science is that in order to propose explanations, one has to know how to do it; that is, scientists need a procedure of giving an explanation that is suitable for the given phenomenon, just as they need a procedure suitable for formulating a statistical test for a certain hypothesis. For example, the t-test (and the statistical inference) details a clear procedure for how to test a hypothesis concerning the difference between means. This procedure details a way of calculating that depends on whether the means are linked to two independent samples or if the standard deviation is to be calculated on the basis of one sample. Just as justification for the use of the t-test is based on several statistical assumptions and mathematical developments, so the procedure for giving scientific explanations is justified by various considerations—logical, epistemological, ontological, and so on up to commonsense considerations of daily life (these justifications are described in chapters 2 and 3, which review different models of explanation). In other words, my treatment of the model of scientific explanation is not like that for a theory of explanation, but like that for a procedure for proposing explanations, when the justification for this procedure is principally theoretical. From this angle it arises that an explanation model may not be tested empirically because this model is not a kind of theory (which is subject to an empirical test) but a procedure justified by theoretical considerations and the degree of match of the procedure to realization of its purpose in the field under discussion.

2. *Explanatory theory (or explanatory hypothesis, model, law, mechanism) in itself is not able to propose an explanation.* To explain a phenomenon, suitable explanatory information is needed. In science this information acquires the form of theory, hypothesis, model, and so on. The theory in itself is not able to explain the occurrence of the studied phenomenon. For the theory to do this, it has to function in the framework of a procedure suited to giving scientific explanations.
3. *The functioning of a procedure for an explanation is based on rational relations.* Rational relations include different kinds of relation, from logical, mathematical, statistical, and causal to practical. To propose a scientific explanation based on nonrational relations is impossible because what one will get is fortuity and chaos.
4. *A procedure of explanation is an integral part of the methodology of science and is closely connected especially to a procedure of the empirical test.* The methodology of science is a collection of methods/procedures for conducting research whose main purpose is to bring about understanding of the world and the creatures that have developed in it (i.e., humans, other animals, and plants). Whoever follows the explanation procedure will reasonably reach understanding of the given phenomenon. (The chief difference between explanation and understanding is that understanding depends on a person's awareness and that there are several understanding procedures that differ from those accepted in science as explanation procedures. (See chapter 1.) Explanation procedures are indissolubly bound to a procedure for empirical testing, such as the Hypothetico-Deductive (H-D) method (see Hempel, 1965, 1966). Without the explanation provided by a given theory passing an empirical test at some time, that explanation is liable not to be accepted by science and to be considered an ad hoc explanation (i.e., an explanation tailored to the specific phenomenon alone). In other words, without an empirical test, we will never be able to know if the given theory is false, and as a result all the pillars of science tremble. In addition, it has to be stressed that from the methodology of science we will never know with absolute certainty that the given theory is entirely correct (despite the many tens and hundreds of empirical attestations supporting it). Scientific research never ends, and anyone who believes that a certain theory is wholly correct in fact drops out of the empirical science game (see Popper, 1972).
5. *The scientific explanation is a kind of description that attempts to approximate the cryptic process that it was responsible for, involved in—the occurrence of the studied phenomenon.* And in slightly greater detail, this description of the explanation in science is set on the following three basic assumptions:

1. There exists in nature some kind of Unknown Real-Process (URP) that is responsible for the occurrence of the studied phenomenon under certain conditions (except for the assumption regarding reality, the rest is unknown);
2. Science proposes a theory (hypothesis, model, mechanism) that tries to approximate the URP and its function (the ways of its action in the context of the studied phenomenon under certain conditions), thereby giving a reasonable explanation for the studied phenomenon;
3. The proposed theory is based on the scientific knowledge collected in the area under discussion, and the degree of proximity of this theory to the URP is measured based on the degree of success/failure of this theory.

This is one of the most important ideas underlying this book. Chapter 1 and chapter 7 develop the idea in several directions. I do not want to overdo the issue, because the idea in essence seems simple and correct, and also its way of presentation here and in the two chapters is direct and clear. The approach here (and of course in the entire book) is realistic in the broad sense. That is, I perceive science as a system that attempts to answer the question of how nature (inanimate, plant, and animal) functions, that is, a system that attempts to decipher nature's code: how it is built, how it works, and according to what. And to answer these questions we must of course assume that nature does not behave randomly or chaotically, and what takes place in it attests to order, to some hidden processes that are responsible for the phenomena that we observe under certain conditions.

6. *To date no one has been able to develop a theory that will describe and explain the connection between neurophysiological activity in the brain and consciousness (i.e., our awareness of the world and ourselves, of whatever reaches us through our senses [directed inward or outward])—akin to what we have succeeded in doing in the sciences;* examples are the development of a theory that explains the connection between mechanical energy and heat, between mass and energy, between one kind of matter and another. In other words, to date no one has succeeded in developing a mind/body theory. This fact has an important implication for the explanation of behavior, because almost all publicly observable behavior is replete with consciousness. And in somewhat more detail:

   1. It is not possible to explain behavior solely by giving mechanistic explanations (explanations that are accepted in the sciences, in behaviorism, in cognitive psychology, and in physiological psychology),

because this explanation will be only partial. It neglects the all-important aspect of consciousness;

2. To propose a better explanation of behavior, account must be taken of the conscious aspect, but because the prevalent methodology in psychology is directed at treatment of mechanistic phenomena alone, and a mind/body theory has still not been developed, at present it is not possible to suggest an appropriate explanation of behavior;
3. It will be possible to propose an improved explanation for behavior given the following conditions:

   a. A procedure must be developed for proposing mentalistic explanations (i.e., explanations set on the individual's inner world, such as will and belief);
   b. It has to be shown that a mentalistic explanation meets the accepted methodological requirements of science;
   c. A theoretical framework has to be developed that allows coherent use of these two kinds of explanation (mechanistic and mentalistic). This account will be better than a purely mechanistic explanation.

This is one of the most important ideas sustaining the present book. Chapter 5 summarizes the empirical–theoretical infrastructure wherein the mind/body theory has not yet been developed, and chapter 6 describes in detail (a) Methodological Dualism, which shows that the mentalistic explanation also meets the methodological requirements of science like the mechanistic explanation, and (b) the Multi-Explanation Framework, which allows one to construct a coherent theory in any psychological domain based on these two kinds of explanation: the mechanistic and the mentalistic. The important points emphasized in the present approach are as follows.

First, the goal of the book is not to offer another solution to the mind/body or consciousness/brain problem. Quite the reverse: the goal is to circumvent this problem, thereby to achieve another goal, but no less important, namely to enhance the explanation of conscious behavior.

Secondly, the present approach does not contend with the consciousness aspect of behavior outside the game rules of accepted science (e.g., by inventing new, extra-scientific, game rules, or drawing a boundary between the humanities and the sciences, as in the hermeneutic approach). The effort takes place within the framework of science by constructing an argument showing that the mentalistic explanation can be encompassed by the accepted methodological framework of science.

Thirdly, the theoretical structure developed according to idea number 6 (above) also accords with the general approach to explanation of the behavior of the individual proposed in idea number 5.

ORGANIZATION OF THE BOOK AND BRIEF SUMMARY OF ITS CHAPTERS

While the book's seven chapters plot an ideational course beginning with chapter 1, on the relation between explanation and understanding, and ending with chapter 7, on a general procedure for giving explanations for the behavior of the individual, the chapters are written in such a way that each can stand as an independent unit.

Chapter 1 discusses the relation of explanation and understanding. In 1948, the year Hempel and Oppenheim published the foundational paper on the structure of explanation, the view prevailed that scientific understanding is based on scientific explanation. About fifty years on, the literature began to indicate new approaches suggesting that scientific understanding is not entirely built on explanation, and that understanding has additional sources such as the cognitive ability of the scientist herself and the degree of use she makes of the explanatory theory. An article by Lipton (2009) propelled these notions to the extreme, and raised arguments that support the idea of understanding without explanation. This article sparked several debates in the literature and moved me to propose a relatively new approach to the connection between explanation and understanding, namely the Dual Theory of Understanding. This theory, which posits a certain nondependence between explanation and understanding, is based on two major key points.

First, scientific understanding is based on consciousness of the scientific content, that is, of the phenomenon under discussion, the explanation, and the scientific knowledge associated with them. Because awareness of this scientific understanding is the birthright of the human being exclusively, it is possible to imagine that a sophisticated robot would supply an explanation for a given phenomenon without understanding the significance of the explanation. So here arises an argument that supports separation of explanation and understanding: while giving an explanation is a mechanical, rational process, understanding the explanation requires a human's consciousness.

Secondly, even before empirical scientific research began to focus on scientific explanation (in 1948), the human being tried to understand the world and himself by means of certain procedures that foster understanding of the phenomena of interest. The chapter sets forth several illustrations of procedures for understanding of this kind: a procedure that allows understanding of a certain phenomenon on the basis of a match between it and some scheme that exists in the individual's cognitive system. For example, for one person a certain shape in the clouds has the form of a bear, for another the form of a hyena; understanding perception of a certain phenomenon in person A is possible if person B gets into person A's shoes—that is, B has to grasp the given phenomenon from A's viewpoint. Understanding a phenomenon comes about through its representation visually; the passage of a square through a round hole becomes possible when the diagonal of the square is smaller than the diameter of the circle. These procedures are based (a) on the underlying assumption that there is some Unknown Real-Process that is responsible for the appearance of the non-understandable phenomenon, and (b) on the notion that understanding arises with the aid of procedures that exemplify the phenomenon over and over from different standpoints, and that these procedures steadily dispel the mist shrouding the Unknown Real-Process. The chapter also discusses differences between understanding and explanation, and suggests that contrary to understanding, explanation is interwoven with scientific knowledge based on the development of a rational scientific methodology.

Lastly the chapter discusses the relation of understanding, explanation, and the development of the scientific model. As an example of the great importance of the scientific model, the chapter considers the model developed by Rakover & Cahlon (1989) called the Catch model. It is intended to reproduce from a witness's memory a picture of the accused's face. The phenomenon whose understanding is being sought is presented as a model of reality, and what explains it is itself a certain theoretical model (which highlights some points and hides others) and that understanding the phenomenon becomes possible as a result not only of the model's success but also (at times mainly) of its failure to propose good explanations (predictions)—a situation that encourages innovative developments.

Chapter 2 reviews critically nine models of explanation discussed in the philosophical literature that are based on research in the sciences. The professional literature lists a larger number of models, but I decided to consider those reviewed here for several reasons, the most important being these: my impression is that these models (a) exerted great influence on the professional literature—that is, on the one hand, they aroused great criticism and on the other hand, were a standard for imitation; (b) are practically applicable in research itself in a relatively straightforward way; and (c) are likely to have a great effect on giving explanations in psychology. Take, for example, Hempel's (1965) model: almost every study in psychology uses the basic ideas of these models according to which an explanation is linked to a prediction that arises from the theory and certain real conditions, and that matches the research results.

The chapter opens with a review and discussion of the three models proposed by Hempel: the Deductive-Nomological (D-N) model, the Deductive-Statistical (D-S) model, and the Inductive-Statistical (I-S) model. Together these are known as the covering-law theory because the common idea is that a determinist law, a statistical law, or a statistical generalization cover explanatorily the studied phenomenon. In the first model, the D-N, the cover is accomplished by means of a deductive argument—so also the second model, the D-S (by means of a deductive argument), and the third model, the I-S, by means of an inductive argument. The three models gave rise to fierce criticism, as a result of which it became clear to the researchers that not only were these models not suitable for all fields of research, such as psychology and history, but also they raised difficult problems in fields of their application in the sciences. Ultimately, these models do not constitute either a sufficient or a necessary condition for giving a scientific explanation. As a result of this criticism, other researchers proposed alternative models of explanation.

The first alternative model that the chapter reviews is the Statistical-Relevance (S-R) model proposed by Salmon, Jeffrey, and Greeno (1971) and Salmon (1990). The important ideas behind the development of this model are, first, that the explanation need not necessarily rest on a logical argument. Salmon bases this model on a comparison of two probabilities: a posteriori and a priori. Secondly, this model need not necessarily be of high probabilities, close to 1, as Hempel's I-S model requires. This model similarly experienced sharp critique, and eventually Salmon himself abandoned it for a new explanatory model based on the concept of causality. This concept was the source of several explanation models discussed in this chapter.

The first causal model that the chapter reviews is the Causal-Mechanical (C-M) model proposed by Salmon (1984). The basic idea on which this model rests is the view that the world is a network of interactive causal processes. The explanation places the studied phenomenon within this network, and the causal process spreads through space and time. The second causal model is the Manipulationist-Interventionist (M-I) model suggested by Woodward (2003). The basic idea on which this model rests is the causal connection between X and Y through a certain manipulation that changes X; hence Y changes accordingly and systematically, and thus the causal explanation allows seeing what is likely to happen as a result of the counterfactual operation of a certain manipulation. The third explanatory model is the Kairetic model proposed by Strevens (2008b). On the assumption that every studied phenomenon has several causes, this model tries to develop a method that will make it possible to find the most important causal factor that creates a difference, by a process of elimination. One by one, factors are removed from the list of causes potentially responsible for the studied phenomenon, and a test is done for whether the phenomenon has disappeared (or is likely to disappear). This test is repeated until only the factors (or factor) are (is) left without which the given phenomenon would not have occurred.

The chapter concludes with a discussion of two models. One is the Unificationist model developed by Kitcher (1989); the other is the Pragmatic model proposed by van Fraassen (1980). The former posits that understanding the world depends on the ability to derive from the same patterns of inference many phenomena that are observed in scientific research. One of the important examples supporting this approach is Newtonian theory that united explanatorily a collection of separate phenomena through reuse of a limited number of laws and calculus. Van Fraassen's Pragmatic model turns attention to the practical aspects, the background factors, of explanation as giving an answer to the question: subject of the question, the contrast-class (the possible alternatives for giving an answer), and the relevance-relation, a kind of relation between giving the answer and the question (e.g., "Compared with the alternatives of the answer, the phenomenon occurred because . . .").

An important conclusion that arises from this review of the models is that there is no single concept for scientific explanation, and it is reasonable to suppose that each of the models reviewed highlights one or two aspects of the notion of explanation, which is so complex and multidimensional.

Chapter 3 studies critically three models of explanation: rule-explanation; functional, teleological models of explanation; and a new-mechanistic model of explanation. These models were developed on the basis of research in the sciences and human behavior (and the behavior of supreme animals). Despite this, it is worth stressing already here that these models, too, despite their improving the explanation for human behavior, are unable to handle the most important component of this behavior, namely conscious experience. My proposal how to do this, that is, how to insert the mentalistic explanation under the wings of the methodology accepted in science, appears in the last three chapters of the book.

Chapter 3 points out that it is not possible to apply any of the models reviewed in the foregoing chapter to human behavior, for example, in the following cases: Dan stopped his car when the traffic light changed to red, or Dan stood on a chair in order to take a book down from the shelf, because for these kinds of behavior there exists no law, theory, or suitable empirical generalization. For example, stopping at a red light is not explained by an appeal to a law or empirical generalization that covers behavior of this kind, but by looking up a traffic rule created by a human being and that a driver is obliged to obey. The chapter discusses seven important differences between behavioral rules and laws or theories: for example, laws are universal and rules are local; laws cannot be changed arbitrarily, but rules can; and laws are subject to an empirical test, and rules are not (rules are evaluated according to the measure of their meeting the purpose for which they were devised). Although not all rules are present in consciousness (e.g., the elementary rules for creating language, which are innate), they differ from laws of nature in that they concern information.

Explanation by rules is also different from teleological explanation. For example, people are not always aware of the aim of the rules that they obey automatically without thinking. The discussion of functional, teleological explanation treats two approaches that attempt to characterize these explanations in causal terms. The approach developed by Nagel (1961) tries to answer the question of *why* a certain component appears in a given system; Cummins's (1975, 1983) approach tries to answer the question of *how* a certain component succeeds in contributing causally to the efficient function of a given system. Important and interesting criticism was leveled against these two approaches to functional explanation.

The chapter ends with a discussion of the new-mechanistic explanatory model. According to this model, an explanation for the given phenomenon is to be proposed by an appeal to a mechanism that has to be decomposed into different components with different functions, unique in their array and interaction, which produce the studied phenomenon. One of the most important researchers who developed this approach, Bechtel (2008a, 2008b, 2008c), draws a distinction between a new-mechanistic explanatory model, which suits the sciences, and one that suits psychology, that is, a mentalistic mechanism, which is meant to process information—the Information Processing (IP) mechanism. To illustrate this mechanism, chapter 3 analyzes the theoretical model for an explanation of the Face-Inversion Effect (FIE) (see Rakover, 2013). According to Rakover's theory, in the individual's cognitive system a scheme develops for perceiving faces upright, which processes faces that are presented inverted, and explains most of the phenomena connected to FIE. The chapter notes that there is an important difference between Bechtel's (2008b, 2008c) approach to a mentalistic mechanism and Rakover's approach, which is developed in chapters 5–7. While according to Bechtel processing information in a mentalistic mechanism occurs on the nonconscious cognitive level, according to Rakover mentalistic activation takes place on the conscious level.

Chapter 4 discusses whether the explanation model in itself is subject to affirmation/refutation through empirical observations and what are the principles underlying the construction of models for giving explanations in science and in psychology. It emerged that a significant share of the critiques leveled against various explanation models were based on known observations that had an accepted explanation, whereby they showed either that the model was too broad because it contained manifestly unreasonable explanations or that it was too narrow because it did not refer to explanations clearly acceptable in science.

Chapter 4 raises arguments against the idea that it is possible to test the degree of correctness of the explanation model by empirical observations. Two important arguments are these. First, the prediction that emerges from a certain theory under certain conditions is likely to suit or not to suit the given phenomenon. In both possible cases—of confirmation or falsification—the empirical result carries no implications for the explanatory model itself, but only for the theory placed in the explanatory model. In essence, the source for this argument is the view that the explanation model is a procedure that guides the researcher on how to propose scientific explanations. An explanation model is not like a hypothesis or theory (or theoretical scientific model), which may pass/fail an empirical test. A procedure is assessed by its rational course of action (e.g., it does not create situations that contradict each other) and is suitable for application in a given field. In this sense, a procedure for giving explanations is akin to an administrative procedure, such as traffic rules that advise people how to behave in certain situations. A rule of the road is assessed as suitable or unsuitable for a certain field of behavior, but it is not judged right or wrong. (For example, the rule "Stop at a red light" is judged unsuitable for a meeting discussing faculty promotions at Haifa University.) The chapter goes on to contrast this distinction to the accepted distinction between descriptive, normative theories and prescriptive ones.

Secondly, we may look at the argument that a low amount of use attests to the wrongness of an explanatory model M_x. Here the question arises how we may explain this finding of dearth of use of model M_x. Clearly, it cannot be explained by M_x itself (because it has failed), so we must seek another model M_y! However, it is reasonable to suppose (according to those who assume that an explanation model's rightness/wrongness is empirically testable) that this model too fails empirically. Why? Because there is very high probability that very often M_y proposed predictions that failed. Hence, not only is M_y not suitable for use as a procedure for giving explanations, but also no explanation model exists that can explain the finding of slight use of M_x—because all models for giving an explanation have at some time failed the empirical test (in the end, the rightness of any scientific theory is temporary!). Clearly then, success/failure has no implications for the degree of correctness of a procedure for explanation.

Based on a review of the explanation models found in the literature, chapter 4 offers a list of several constraints, characteristics, and properties for constructing an explanation model, a significant number of which have been described above, and these make up the infrastructure for building explanation models for psychology (see especially chapters 6 and 7).

Chapter 5 discusses the following fundamental question: Is it possible to comprehend mentalistic concepts and explanations (based on the individual's inner world) with the aid of mechanistic concepts and explanations (based on and accepted in the sciences)? The answer is negative, and it forges the way to the development of a new methodological approach to proposing explanations, namely the Methodological Dualism (MD) and Multi-Explanation Framework (MEF), whose principal purpose is to propose a procedure to explain behavior that is brim-full of consciousness (see chapter 6). This methodology is based on the theoretical–empirical fact that as of today no one has succeeded in developing a mind/body theory. Were it possible to propose mind/body theory, it would be possible to forgo the MD and MEF approach, because everything would obtain a mechanistic explanation. But as chapter 5 shows, this is not the situation so far.

Chapter 5 breaks down the above fundamental question into the following four questions:

1. Can human behavior be explained mechanistically alone, without recourse to the concept of conscious experience?
2. Has a theory been successfully developed that explains the connection between neurophysiological processes in the brain and conscious experience?
3. Can mentalistic explanations anchored to the concept of conscious experience be reduced to mechanistic explanations (e.g., to a neurophysiological explanation)?
4. Has a very complex and sophisticated computer proven able, like us, to behave with conscious experience?

An affirmative answer to any of these four questions means that there is no need to develop MD and MEF because the standard methodology developed in the sciences can handle research in psychology. Only a negative answer to all these questions will oblige one to think about the development of a new explanatory approach suited to the field of psychological research.

The answer to the first question is negative. Not only is there an immense collection of behaviors by human beings (and by supreme animals) that require mentalistic explanations beyond mechanistic, but also it is hard to accept that the mentalistic explanation is a kind of epiphenomenon. Furthermore, the purely mechanistic explanation is in fact right for robots or zombies devoid of consciousness.

The answer to the second question is negative. A philosophical approach or scientific theory has yet to be found that explains how a neurophysiological process in the brain becomes or creates consciousness.

The answer to the third question is also negative. It arises that mentalistic theories cannot be founded on mechanistic theories. One of the important factors that prevents this kind of reduction is the multiple realizations thesis. If a mental situation or process is defined by its functioning, it may be realized in various ways and with various materials. Hence, if reduction of a mentalistic theory to a mechanistic one depends on the possibility of a monovalent identification of concepts in one theory with another, this possibility is destroyed because according to multiple realizations there are many ways to realize and identify mentalistic concepts that appear in a given theory.

The answer to the fourth question is negative also. It arises that the computer, as sophisticated as it may be, is incapable of reaching a condition in which it will undergo like us humans a conscious experience. One of the most powerful and famous arguments in support of the negative answer is Searle's (1980) thought experiment. According to this, activation of physical signs by means of syntactic rules cannot impart semantic meaning to these signs.

Finally, the chapter attempts to summarize the main obstacles facing the development of a mind/body theory, such as consciousness epiphenomenalism. This arises from the fact that the entire explanation function of behavior falls on the shoulders of the neurophysiological process, and on the inability to measure mental concepts in the same way as material concepts.

Chapter 6 describes and justifies the methodological approach for proposing explanations titled the Methodological Dualism and Multi-Explanation Framework. As noted, this approach is based on chapter 5, which supports it theoretically and empirically in that no one has as yet succeeded in developing a mind/body theory. Chapter 6 is divided into four sections. The first proposes an argument justifying, on the basis of the results of chapter 5, why the MD and MEF should be developed. According to this argument, if we attempt to offer an improved explanation for behavior filled with consciousness, there is no other way than adding to the mechanistic explanation the mentalistic explanation. This leads to the development of the present methodological approach.

The second part of the chapter develops Methodological Dualism. First the chapter demarcates the boundaries between the mechanistic and the mentalistic explanation, and focuses particularly on the teleological explanation that combines the individual's will, belief, and behavior that realizes his will, and suggests conceiving the following as a mentalistic explanation model, a procedure for creating specific will/belief explanations:

[Will/Belief]: *If X wants G and believes that behavior B will realize his/her will, then X will perform B.*

This is a new proposition, and it is central to the present approach. It immediately raises the following question: Does this explanation procedure satisfy the methodological requirements for explanation accepted in science? The answer is yes.

The third section develops a theoretical framework, the Multi-Explanation Framework, by means of which it will be possible to construct a specific theory (an MEF theory) for a specific research subject based on two kinds of explanation: mechanistic and mentalistic. Because the theory rests on multiple explanation models, a major difference arises in the way the theory supplies an explanation in science and the way an MEF theory supplies an explanation in psychology.

An MEF theory is based on a match between explanation procedures (mechanistic or mentalistic) and behavior (we may call it the "explanation/behavior match," for short). While there are behaviors that can be accounted for satisfactorily by an appeal to mechanistic explanations, most behaviors and their decomposed components must be approached by both mechanistic and mentalistic explanations. The coherence of a specific theory based on the MEF is achieved by matching appropriately the explanation procedure to the behavior and its components by means of several guidelines based on theoretical–empirical knowledge and the principle of explanation matching. According to the latter, the components of a mechanistic behavior (a whole behavior that has been accounted for mechanistically) cannot take mentalist explanations, while the components of a mentalistic behavior (a whole behavior that has been accounted for mentalistically) can take mentalistic as well as mechanistic explanations. The appropriate explanation/behavior match solves several methodological problems raised by an MEF theory.

The last section of chapter 6 summarizes the main ideas of the present approach and compares it with other relevant approaches.

Chapter 7 is titled "A Proposed General Explanation Procedure (GEP) for Understanding an Individual's Behavior." My impression from the history of psychology as a science is that this scientific branch used and uses a fairly large number of explanation models, schemes, or procedures, for example, use of different variations of Hempel's models, different kinds of causal processes, mechanisms based on the workings of the computer, and different kinds of procedures based on neurophysiological processes. This multiplicity of explanation models (including the present proposal of a mentalistic explanation model, [Will/Belief], and the proposition to construct a theory based on MEF described above) raised the following question: Is it possible to propose for psychology a general procedure for giving an explanation that is likely to spread its wings over other models too? The answer is affirmative, and is detailed in chapter 7. It is based on three major theoretical ideas:

1. It is assumed that in the individual a certain real-process exists that is involved in the generation of the behavior under study; although the nature of this process is unknown, it is further assumed that it presents the real explanation for the behavior in question (this process is called the Unknown Real-Process);
2. A researcher may approximate the URP by constructing a theory (or a hypothesis, model, mechanism, etc.) that portrays the properties and operation of the URP in generating the studied behavior. Because of this proximity, it may be proposed that the theory provides us with a good account of the behavior in question;
3. The degree of proximity of the theory to the real-process is estimated by a "proximity index," based on the theory's successful predictions. A confirmed theory is closer than a refuted theory to the URP and the real explanation.

The GEP is a general procedure for giving an explanation in two respects. First, the assumed real-process can be approximated by diverse theories or hypotheses, models, and mechanisms (here called "theoretical-structures").

Secondly, the GEP may handle different kinds of explanation procedures (schemes, models) that researchers have been using in their studies. These models may be viewed as proposals for the different attributes of the Unknown Real-Process and the way it functions.

To illustrate the GEP, the chapter analyzes several explanations given for learning in animals, perception and recognition of faces, and everyday behavior connected with purposive explanations (will/belief accounts). In addition, the chapter discusses the basic philosophical concepts and assumptions that underlie the GEP.

Chapter 1

Understanding and Explanation

In their first year of psychology studies, students are told that the aims of science are in essence the construction of theories that provide understanding, predictions, and control of variables responsible for the appearance of the studied phenomenon; these variables are operable, and therefore reality, nature, the world, can be changed.

This section, as can be seen, is very short, and I assume that the reader understands the fundamental meanings of the concepts that appear in it, such as theory, explanation, understanding, predictions, and control. Still, I would like to discuss here whether there is a difference between understanding and explanation. But before this, I must distinguish the everyday meaning of understanding and explanation from their meaning in science.

We may begin with an example that is a variation on an actual event that happened some years ago in a town near the city where I live. In a horrendous road accident involving a school bus ferrying kids of the "Carmel" (pseudonym) school and a train, twenty-three children were killed and fourteen were seriously injured. Anyone reading about this awful event in the newspaper would almost automatically ask questions: Why did it happen? How did it happen? Who's to blame for this shocking accident? The next day the paper published a short interview with the rabbi of the town. He explains to readers the reason for the dreadful event: it emerges that the small prayer scroll, the *mezuzah*, nailed to the doorpost of Carmel school, was flawed! For the believer, this explanation settles the matter because he understands that a great sin was committed here, a serious transgression, a sacrilege, which God avenged with cruel punishment. This explanation is of the kind acceptable only to people unwaveringly faithful to God with all their heart and who observe all the religious commandments. These people have internalized deep down as an immutable law this rule: straying from religious obedience entails mighty punishment from God. This rule explains punishment for any deviation from religious behavior as laid down in the religion's commandments; for example, it answers the chilling question why six million Jews were systematically annihilated by the Nazis. The answer is because the Jews sinned before God.

These are not explanations that science is willing to accept because they do not uphold the game rules of science, namely scientific methodology; they do not withstand rational criticism. Returning to the dreadful schoolbus accident, it arises that the school was established eight years before that event. That is, for eight years a flawed and profane *mezuzah* was nailed to the Carmel entrance, and nothing terrible happened! How may this be explained? Among the six million Jews slaughtered in the Holocaust, there were thousands of unborn babies in their mothers' wombs, thousands of infants who had not yet learned to walk, thousands not yet of kindergarten age, thousands of children still unblemished by the commission of any iniquity, any transgression, thousands of youngsters still too young to be liable to sin—all murdered innocent of any wrongdoing! How is this explainable?

So I believe that this point is clear: I don't plan to treat any explanation and understanding except by way of scientific explanation and scientific understanding, that is, by way of explanation and understanding that meet the requirements of accepted scientific methodology; by this term I mean in the most general way all the principles and rational work methods on which empirical science runs. These principles and methods, and the rational basis on which they rest, are discussed in the following.

A SHORT SURVEY OF THE SCIENTIFIC STATUS OF THE TERM *UNDERSTANDING* AND ITS RELATION TO THE TERM *EXPLANATION*

This survey is principally based on the book *Scientific Understanding: Philosophical Perspectives*, edited by de Regt, Leonelli, and Eigner (2009b). It is a collection of articles on scientific understanding and explanation. It may be said that following the publication of Hempel and Oppenheim's (1948) article on studies in the logic of explanation, philosophical study focused on explanation, not understanding. After an interval of about fifty years, philosophers began systematically to discuss the subject of understanding. The perception of explanation that Hempel and Oppenheim took as their basis was the philosophical perception accepted at that time, namely the positivist approach that prevailed in the 1940s. This regarded the development of logical models as providing an explanation for one of the most important goals of understanding the development of science in all its branches. Hempel's (1965) book pursued this perception, and proposed models not only for determinist occurrences of events, but also for probabilistic phenomena—a general approach called "covering law theory," because Hempel's models explain the given empirical phenomenon by showing logically that it was covered by (contained under) scientific laws (determinist, probabilistic, and well-established empirical generalizations).

The explanation models proposed by Hempel were contained within a general framework of the objective scientific game rules laid down by the positivist approach; understanding, however, was deemed a subjective-psychological phenomenon ("Aha—I understand why this happened"), which therefore had no place in the scientific framework (see also Trout, 2002). In response to Scriven's critique of Hempel, the latter writes: "Indeed, such expressions as 'realm of understanding' and 'comprehensible' do not belong to the vocabulary of logic, for they refer to the psychological or pragmatic aspects of explanation" (Hempel, 1965, p. 413).

The philosophers de Regt, Leonelli, and Eigner (2009a) write that by this approach, understanding was perceived as a cognitive/emotional response occurring after an explanation was given: "Understanding has typically been viewed as a by-product of scientific explanation" (p. 7). Accordingly, diverse researchers have suggested that scientific understanding arises following an explanation that shows how the studied phenomenon comes under the aegis of a theory or law (e.g., Hempel, 1965); fits into a single theoretical framework (e.g., Friedman, 1974; Kitcher, 1989); and is based on the description of a causal mechanism (a chain or a network of causal factors that act on each other) that fosters understanding how the studied phenomenon was created (e.g., Salmon, 1984, 1990). Here is a very simple example of understanding in consequence of an explanation.

David says, "I don't understand why this stone, which fell from the roof of the tall building of height *a*, hit the ground after a second and a half."

"The explanation is perfectly simple," says the physics teacher, "and is based on Galileo's Law. From this law, based on the earth's gravity, it follows mathematically that if the height of the building is *a*, the duration of the stone's (actually any stone's) fall must be a second and a half."

"Aha," says David, "now I understand why this happened: it's because of the law of gravity, whose operation finds exact, mathematical, expression in Galileo's Law."

This perception was opposed by the researchers de Regt, Leonelli, and Eigner (2009a, 2009b), who contributed articles to the book they edited. They argued that scientific understanding is much more than a by-product of giving a good scientific explanation because understanding the world does not rest only on the relation between the explanatory information, the *explanans*, and the phenomenon to be explained, the *explanandum*, as suggested, for example, by Hempel in his explanation models, but also on the skills of the scientist himself. de Regt, Leonelli, and Eigner (2009a) write, "In other words, while explanation may be viewed as a two-term relation between *explanans* (for example, a theory) and *explanandum* (the phenomenon), understanding is always a three-term relation involving *explanans*, *explanandum*, and a subject" (p. 3).

This distinction between explanation and understanding cuts across the various scientific disciplines, and also across the distinction between explanation and understanding made at the turn of the twentieth century by German philosophers (e.g., Dilthey). They proposed a very interesting methodological dichotomy of the natural sciences as against the humanities and social sciences: the goal of the former was to propose explanations (*Erklären*) for the studied phenomena, while the purpose of the latter was to propose understanding (*Verstehen*) of human deeds and culture. Here the term *understanding* is hermeneutic: a term aiming to express the meaning of a person's activity in the society in which she lives, a term that directs the researcher to reach a hermeneutic understanding of the human being (now and in the past) not from her viewpoint but from the other's. This approach runs counter to positivism (for reviews, see von Wright, 1971; Mahajan, 2011). A group of philosophers in 2009 (those who contributed to de Regt, Leonelli, and Eigner's 2009b book) do not accept this methodological split into the natural sciences versus the humanities and social sciences. Yet they are unwilling to accept the objectivist perception preached by researchers raised on the positivist approach, holding, as stated, that understanding, which is nothing other than a subjective concept, is a by-product of explanation and has no place in science.

de Regt (2009) distinguishes three basic concepts (referring to cognitive conditions) associated with understanding and explanation of a given phenomenon by means of a suitable theory:

1. A sense of understanding: the psychological feeling of "Aha, I understand the thing"
2. Understanding the theory: the scientist's *skill* to *use* the theory
3. Understanding the phenomenon: possessing a suitable explanation for the given phenomenon.

de Regt's foremost argument is that (1) a sense of understanding does not play an important part in understanding and explaining the phenomenon. (In fact, the positions on a feeling of understanding expressed by contributors to de Regt, Leonelli, and Eigner's book range from regarding it as meaningless or harmful to something that could well be useful.) By contrast, de Regt holds that (3) understanding the phenomenon necessarily depended on

(2) understanding the theory, which requires the intervention of the scientist himself, his *skill* to *use* the theory. And because this last component (skill to use) is not entirely objective, understanding is not entirely objective either, even though using theory fulfills a very considerable epistemic role in understanding the phenomenon: it is an essential element in the scientist's ability to understand the theory that provides an explanation for the given phenomenon. He goes on to state that the relation between the *explanans* and the *explanandum* is necessarily linked to the scientist's skills in using the theory but also to his powers of judgment (e.g., the ability to judge what is the most suitable and effective theoretical approach in a given empirical situation).

In addition, de Regt proposes that understanding rests on the scientist's ability to construct models representing the given phenomenon. Resorting to the relevant literature, he posits that a model is a unique description, an idealization of reality (e.g., planet earth, the sun, and other stars are represented in Newton's method as dots possessing different values of mass). The model of the phenomenon is built by the scientist in such a way as to allow efficient application of the theory to the model for the formulation of precise predictions, testable by their comparison with empirical measurements in reality. Thus the scientist has to possess great skill, a creative imagination, and robust powers of judgment in order to construct a model of the phenomenon, which, on the one hand, will be close to reality itself and, on the other, will allow direct application of the theory to this model.

On these grounds de Regt submits that scientists customarily use the same scientific theory that admits the quality of intelligibility, namely a theory that can be credited with an array of advantages that facilitate the construction of scientific models whereby explanations can be proposed for the given phenomena. Practical understanding of the theory, therefore, is nothing but a precondition for scientific understanding of the theory.

de Regt's approach can be summed up as follows: on the one hand, he accepts the way of Hempel and his followers that understanding is obtained through explanation provided us by the theory. On the other hand, he presents an addendum that runs counter to Hempel's position, stating that understanding the phenomenon depends on understanding the theory, and this understanding in part depends on nonobjective elements linked to the scientist herself, namely to her practical abilities to utilize the theory, to judge it and its application to specific cases, and her skills in building models apt for the creation of good explanations.

While most articles in *Scientific Understanding* assume dependence between scientific understanding and scientific explanation, Lipton (2009) offers a contrary approach, expressed in the title of his contribution: "Understanding Without Explanation." He discusses several examples where the individual acquires understanding without resorting to explanation. For example, one may obtain understanding of environmental information by resorting not to the correct explanation but to other methodological means such as conducting observations, running experiments, and using visual models. As an illustration he writes that he was able to understand an astronomical phenomenon called "apparent retrograde motion" of stars in the sky when visiting a planetarium, where he saw a realization of the phenomenon by means of a model of the solar system. The reverse movement of the stars in the sky is an illusory phenomenon due to observations taken from earth: the observed star (e.g., Saturn) is seen to perform a very strange movement: first it moves forward, then stops, then moves backward, stops again, and resumes its forward motion. The scientific explanation for this phenomenon lies in the differences between the speed of rotation of the earth (from where the observation is made) and that of other planets circling the sun, when their location is set off against the black background of the heavens.

Just as the solar system model helped Lipton understand the apparent retrograde motion, so I have been helped by other physical models to understand other interesting phenomena in astrophysics. For example, it is possible to illustrate the astonishing finding of expansion of the universe, of distancing of all the galaxies from each other, a finding that can be expressed in Hubble's Law by use of a simple physical model: draw dots on a balloon and inflate it. As the balloon steadily expands (i.e., as the universe spreads and expands), so the distances between the dots that stand for galaxies in the universe steadily increase. Here no complicated astrophysical mathematical explanation is needed. A grasp of the phenomenon is immediate, and all can see how it takes place before their eyes. Similarly, the effect of mass on space/time in Einstein's General Theory of Relativity is demonstrable: balls of different weights are placed on a sheet stretched out in the air, above the ground. As the weight of each ball is greater, so the dip under the stretched sheet enlarges (like an inverted funnel). Moreover, if we place small beads on the edge of the sheet, they will roll along the walls of the sheet's concave surface, and their movement will represent the curve in the movement of a light ray passing near stars in space. Can it be said that I fully understand the law and the theory under discussion? The answer is no! I am not a physicist, and I don't know the physics and the mathematics involved in Hubble's Law and in the General Theory of Relativity (where, I admit, it is hard for me to take in the elaborate notion of space/time). Yet despite all, these realizations have succeeded in clarifying for me important and interesting matters highly relevant to these two subjects.

Lipton's article contains an analysis of more examples, which I shall not deal with at the moment. Still, it is noteworthy that this article sparked many responses in the professional literature—philosophical papers arguing against Lipton's concept, and attempting to show that there is no scientific understanding without a scientific explanation. For example, concerning the demonstration of stars' retrograde motion, Strevens (2013) suggests that understanding the phenomenon itself (a presentation of retrograde motion) is different from understanding its occurrence (i.e., it differs from formulating the explanation of a phenomenon as an answer to the question of why it happened).

Regarding each of Lipton's examples supporting understanding without explanation (of the question why), Khalifa (2013) argues that an explanation for the given phenomenon exists that entails greater understanding than that acquired by means that are not a sort of explanation. To elucidate this interesting proposition, let us examine briefly the following example (a variation on Lipton's account). In a fixed boxing match, it was decided that the stronger boxer, Baker, would fake being knocked out in the tenth round. What happened in fact was that by chance his weaker opponent, Abel, landed Baker an almighty uppercut that floored him in the fifth. Now suppose that Dan knows about the deal (and accordingly has placed a very large bet on Abel), but he doesn't know about a knockout in the fifth round. So when he hears that he has won a lot of money, he understands the reason for it (the fixed bout) without actually knowing the true reason (Abel's lucky uppercut). Accordingly, in the spirit of Khalifa's critique, the following may be said.

True enough, our friend Dan has some slight knowledge of what happened. But the true explanation offers greater understanding than Dan's. Why? Because contrary to the understanding that Dan has, the real explanation would be about success in situations known as counterfactuals. Woodward (2003), one of the most important researchers in explanation theory, maintains that a causal theory has to answer questions of this kind: What *would have happened* had the situation been different? In our case the true explanation can answer the question of what would have happened had Abel not succeeded in downing Baker in the fifth. And the answer is that Abel would have won. Hence the true explanation provides greater understanding (it deals with counterfactual situations as well) than the explanation possessed by Dan, whose understanding is restricted to the fixed bout. In other words, contrary to Dan's understanding, the true explanation is able to impart to us understanding of Abel's victory in the factual case (knockout in the fifth round) and also in the counterfactual case: Abel was the winner, in keeping with the fixing deal.

THE DUAL THEORY OF UNDERSTANDING

My impression from reading the literature on the relation of explanation to understanding is that there is a certain independence of explanation from understanding: it is very hard to suggest that understanding is not acquired by means of explanation, but it is undeniable that understanding is helped by other methodological means as well, different from those supplied by explanations. This impression served me as the starting point for the development of a new theory of understanding, called the Dual Theory of Understanding (DTU). By this theory, as we shall see, in a single theoretical framework two perceptions are arranged coherently: understanding rests on explanation, and understanding is also acquirable not by means of explanation. The basic idea underlying DTU is this: just as there are special procedures for proposing scientific explanations, there are different, specialized, procedures for developing understanding, namely understanding-procedures different from explanation-procedures.

Both explanation and understanding have degrees of approximation to what may be termed correct explanation and correct (complete) understanding. The product of explanation (correct, incorrect) and understanding (correct, incorrect) results in four possibilities that may help in understanding the present thorny issue.

Correct Explanation, Correct Understanding

This situation is common. Researchers have proposed a correct explanation, and their understanding of the studied phenomenon is also correct. To illustrate, let us revisit free fall with the aid of Galileo's Law: the explanation as well as understanding of the law (reinforced by Newtonian theory) and of the way the empirical phenomenon is connected to the law through a procedure of deductive explanation are correct. In my view, certain cases that Lipton (2009) discusses belong to the present category: he proposes correct understandings of a phenomenon where in the background lies the correct explanation, for example, the explanation of apparent retrograde motion of heavenly bodies (but, as mentioned above, see the criticism in Strevens, 2013; Khalifa, 2013).

Correct Explanation, Incorrect Understanding

In this category we may place the following case: researchers propose a correct explanation, but other researchers are liable not to understand the explanation properly, regardless of their own belief that they do understand properly both the explanation and the answer to questions as to why and how the studied phenomenon occurred. Another case, which fits this category, deals with the possibility that a computer may suggest a correct explanation without understanding the explanation (this topic will be discussed later).

I maintain that the famous example of Tversky and Kahneman, the case of Linda, belongs to the present category (see Kahneman, 2011). Very briefly, Linda's case is this: first, participants in the experiment were presented with a description of Linda. She is thirty-one years old, single, brilliant, and speaks her innermost thoughts. She has a degree in philosophy, engages heavily in issues of discrimination and social injustice, and has also joined demonstrations against nuclear arms. Next, the participants were asked what seemed more reasonable to them: (1) Linda is a bank clerk, or (2) Linda is a bank clerk and is active in the feminist movement. The great majority of the respondents (including academics) chose the second option, even though statistically the probability of the first was greater than that of the second. Why? Because the second possibility is based on the product of two probabilities (that are less than 1), which is less than the probability of either one of the two probabilities; for example, given that P(clerk) = .3 and P(feminist) = .5, it follows that {P(clerk) × P(feminist)} = .15, which is less than the probability of either of these two options.

From our viewpoint, what we have before us is this. On the one hand, we have a correct explanation, a correct statistical answer to the question about Linda; on the other hand, we have before us an incorrect answer, an understanding based on a heuristic of representation, namely on an incorrect understanding whereby the second possibility better suits Linda (who represents this possibility).

Incorrect Explanation, Correct Understanding

For this category two cases may be suggested. In one, the researchers understand properly the studied phenomenon, but suggest an incorrect explanation. In the other, as we shall see later, one uses certain procedures that substantiate (through the senses, mainly visual) the studied phenomenon and enhance at least a partly correct understanding of the phenomenon.

Incorrect Explanation, Incorrect Understanding

To illustrate, Dan was injured yesterday when he crossed the road. How may we explain this lamentable event? David offers an incorrect explanation: when a black cat crosses a person's path, something bad will happen to that person. Before Dan was run over, a black cat crossed his path (in Dan's neighborhood there are lots of black cats). David goes on to say that he understands what happened to Dan, because once something similar happened to him: a black cat crossed his path and about an hour later he tripped and broke his leg.

In view of the above, when an empirical phenomenon is not understood and requires an answer to questions about it—why and how did it happen, the following may be proposed: when understanding-procedures encompasses the phenomenon, understanding can be achieved. This notion follows in the footsteps of the notions of Schurz and Lambert (1994), who appeal to the researcher's cognitive knowledge: "To *understand* a phenomenon P is to know how P fits into one's background knowledge" (p. 66). The question how knowledge about P fits into scientific knowledge needs a complicated solution, which I shall allude to in the course of this discussion (see an extensive discussion of this question in Schurz and Lambert, 1994).

Scientific Understanding as Awareness of Scientific Contents

I maintain that a robot is able to offer a satisfactory explanation for the given phenomenon without itself understanding it (see the second category above). One who understands the robot's explanation is a human being, because he/she is gifted with an amazing ability to be in a state of consciousness, awareness, which, in my opinion, is a crucial condition for understanding. That is, without awareness there is no understanding, and if there is awareness of a certain phenomenon, there is direct understanding of the fact of the existence of this phenomenon. Awareness induces not only the phenomenon itself, but also the most complex knowledge—for example, the following questions: Is the given phenomenon chance or not? What caused the appearance of the phenomenon? It induces scientific explanation—understanding an explanation for the given phenomenon, incorrectness of the explanation—understanding that the explanation is not correct, and also the knowledge that the explanation is not a scientific explanation but a religious interpretation. All these are contents induced by awareness, contents beyond basic, primal understanding, for example, that this object, this creature, exists in the world in which I live. This approach parallels Penrose's (1994) position proposing firm ties of consciousness, awareness, to understanding and to intelligence: there is no intelligence without understanding, and there is no understanding without awareness; and the approach also parallels Strevens's (2013) opinion: he wrote the following about an observation of a cat on a mat: "You might understand *that* the cat is on the mat, meaning that you are fully aware, you are entirely conscious, of the cat, the mat, and a certain spatial relation between them" (p. 510). In this context Strevens held that this direct understanding is nothing other than the fundamental deep relation between the mind and the world. About this relation he goes on to express the following sentence, with which I wholeheartedly agree: "The question of the nature of this relation is perhaps the deepest in all philosophy" (p. 511).

My impression is that to date no theory has been discovered that connects the mind with the brain—a mind/body theory; not even the most intelligent and sophisticated robot has developed consciousness like that of the human (or of animals such as primates, dolphins, dogs, cats) because those who have created it could not have based it on a mind/body theory. So I am not surprised that to date there are no robots capable of understanding the explanations that they have provided. This view matches Penrose's: "I would maintain that there is yet no physical, biological, or computational theory that comes very close to explaining our consciousness and consequent intelligence, but that should not deter us from striving to search for one" (1994, p. 8).

My approach here also harmonizes with the conclusions arising from one of the most famous thought experiments, Searle's (1980) the Chinese Room. (In fact, in 1999 I published an article supporting wholly the conclusions arising from this experiment.) Very briefly, the Chinese Room experiment describes a super-perfect computer, which we may call China-3, which answers correctly in Chinese any question presented to it in that language. Next the experiment proposes that David, who does not understand a word of Chinese, conduct all the computations of China-3 instead of this computer: he will take the Chinese input, run on it all the computations done previously by China-3, and produce output in Chinese. Chinese experts who judge his answers to the questions will say they are entirely logical and correct. As a result of performing these computational acts, will David understand Chinese and develop awareness and understanding of this language? The answer is a resounding *no*! And sure enough, to this day, computers, advanced as they may be, have not developed awareness, consciousness, akin to that of the human being.

To illustrate this matter—that a robot is able to offer an explanation without understanding it, we may turn to Galileo's Law once more. David asks: "How is it possible to explain that this stone, which fell from the roof of a house of height *a*, hit the ground in a second and a half?" This question is handled by R, a robot specializing in classical physics. This is a most highly perfected robot, in which the Robotica computer company has installed sophisticated state-of-the-art software capable of answering all questions in the field of classical physics taught at high school. R reads the question and answers immediately: "The answer is quite simple, David, and is based on Galileo's Law, namely $S = 1/2GT^2$, where S is the distance of the fall, G is acceleration caused by earth's gravity, and T is the time of the fall. By this law it is easy to calculate the time of the fall when the distance of the fall is *a*. And now, because what is obtained from Galileo's Law—the stone fell in free fall for exactly a second and a half—matches what happened in fact, it arises that Galileo's Law has succeeded in furnishing an explanation, an excellent answer to your questions, David." David scrutinizes R's answer and says, "Aha, I get it. The stone fell for a second and a half on account of the gravitational force, whose action found precise, mathematical, expression in Galileo's Law." And although we human beings understand this explanation, and reckon that R's explanation is a correct and good one, R does not understand its own explanation. Why? Because of the arguments against the assumption that a robot thinks consciously, like a human, and because no mind/body theory has yet been found, a large body of researchers and philosophers, among whom I myself am numbered, are convinced in the deepest possible way that R is nothing but a cleverly made machine, does not understand its own explanation, and is not in a state of consciousness, awareness, as we, humankind, are.

In consequence of the foregoing, I would like to make a primary distinction between explanation and understanding: while explanation can be given by a machine that is not aware of its actions, understanding requires awareness of the explanation and its elements (i.e., awareness of the *explanandum*, awareness of the *explanans*, and awareness of the rational relations that connect the two). Because to the best of my knowledge the human being is a creature in a marvelous state of consciousness and awareness, here I restrict the discussion to the human being alone. (I hold that a few animal species, such as primates, dolphins, dogs, and cats, are also in a state of awareness but at lower levels than the human. See Rakover, 2007.) In this respect, I do not accept the proposition of Schurz and Lambert (1994) that while understanding addresses the *how* question, explanation deals with the *why* question, because these questions underlie the attempt to understand the world today, and even before the start of development of modern empirical science, beginning from the time of Galileo and Newton.

The second distinction I wish to make is between cognitive understanding (awareness of the content of the answer to the questions of why and how the studied event occurred in certain conditions) and the sense of understanding (the feeling of "Aha—got it"). The understanding that I wish to talk about is the initial, cognitive, understanding, because in my opinion the subjective sense of understanding is nothing other than a response to cognitive understanding. As such, the feeling of relief, even exultation—"I got it"—even though it may be perceived as great reinforcement for researchers, does not constitute an important element in the present attempt to grasp the epistemic/methodological relation of understanding to explanation (but see other approaches to the subject that appear in de Regt, Leonelli, and Eigner, 2009b).

Considering the present discussion so far, I shall take scientific understanding to be based on the following three important points:

1. *Multiplicity of awareness*: Multiple awareness of different knowledge contents relevant to the questions of why and how the given phenomenon occurred: awareness of the given phenomenon, awareness of the answers to the questions of why and how the given phenomenon occurred, awareness of the use of means to obtain these answers, and judicious awareness of the question and the answer as to whether the answers to the *why* and *how* questions are good, and accord well with existing relevant knowledge.
2. *Types and degrees of understandings*: I assume that not only are there different kinds of understanding—but there are also different degrees of understanding. For example, regarding first-year psychology students, my understanding of uses of the computer is less than theirs; however, regarding my understanding of the theory underlying the computer, their understanding is less than mine. But regarding bachelor's students in computer sciences, their theoretical knowledge is greater than mine, even though my understanding of the philosophical implications of the computer (especially about the question of awareness and the cognitive system) is greater than theirs.
3. *Understanding is founded on scientific methodology*: Understanding has to be perceived as anchored to the development of science to date, as meeting the requirements of the accepted methodology of science. By the term *scientific methodology*, as stated, I mean all the rational principles and work methods whereby empirical science functions. In this respect it is understandable that nowadays one may suggest that in the eyes of many, a scientific explanation constitutes a condition for understanding the studied phenomenon.

Following the foregoing discussion on understanding, I shall now consider a few examples of procedures to achieve understanding.

Different Procedures for Understanding

The use of scientific procedures to explain why and how a certain phenomenon occurred apparently began with the rise of empirical science in the time of Galileo and Newton (late sixteenth century). However, the first paper to propose a formal theory and procedure for giving scientific explanations, and which initiated an extremely important research field in the philosophy of science, was published by Hempel and Oppenheim centuries later: in 1948. In other words, the use of special procedures to present explanations in science has developed in the course of the last five hundred years, more or less. What did humans do about this matter of understanding before the onset of the era of empirical science? What procedures did they use to furnish an understanding that would satisfy their curiosity? I surmise that prior to this time when researchers and philosophers began to develop the subject of scientific explanation as a way to understand the world (astronomy, the globe, nature, animals, human beings themselves), the individual in everyday life tried to understand the riddle of the world through other procedures altogether, different from those giving the explanations customary today, procedures that provide a person with understanding. (This is a very hard conjecture to support in historical terms, so I leave it here as is, an idea that might be an amusing mind game.)

These statements may well spark some opposition in the reader's heart, because he is likely to recall that already back in the age of Greek philosophy, Aristotle offered four reasons why a phenomenon occurred the way it did: the material reason, concerning change in the constituents of the given phenomenon's matter; the formal reason, concerning the relation of the phenomenon's components; the active reason, concerning the factor responsible for the phenomenon's occurrence, and the teleological reason, concerning the purpose of the studied phenomenon's occurrence. Aristotle exerted enormous influence on philosophical thought until the start of the development of modern science, which was based on empirical methodology, different in many respects from the approach to understanding the world that prevailed previously. Thus, the question that arose within me was not a historical question—How did people try to understand the world in the distant past?—but the following: What are understanding-procedures that do not rest on explanation-procedures likely to look like? (This is a question that in several respects was engendered by Lipton's 2009 article on "Understanding Without Explanation." Here it should be noted also that Psillos [2007], in the framework of a discussion of scientific explanations, also conducted a short historical review of the rudimentary ideas about explanation that appear in several philosophers and researchers from ancient Greece to the appearance of Hempel and Oppenheim's publication in 1948.)

The simplest way to clarify this subject is through several examples that may be classified into two main kinds of understanding-procedures: one, use of cognitive-psychological tools (e.g., schemes and an empathy process), and two, use of realizations (principally visual) enhanced by two important properties. On the one hand, these realizations themselves are phenomena requiring understanding; on the other hand, they contain clues that may lead to the appearance of answers to *why* and *how* questions. As we shall see, these realizations are not based on historical research but on an attempt to answer the above question (What are understanding-procedures likely to look like?). Some of the understanding-procedures described next may well be perceived as modern by nature. There are two reasons for this. First, I have striven to use contemporary situations, which have a fitting explanation, but are treatable by means of understanding-procedures; second, some of these procedures could serve as aids to presenting explanations that are accepted by modern science.

Understanding by Using Schemes and Prior Substantiated Knowledge

Assume that Umlek from the North Pole and Bugamba from Africa are looking at the sky and Umlek says, "Look, my good friend Bugamba, see—the cloud on the right is the great polar bear that devoured two children of my tribe." Bugamba looks and looks intently, and at last says, "What are you talking about, my good friend Umlek, a bear? There is no such thing as a bear—I have not seen or heard of any such animal. It may just be possible to see, with enormous effort, some kind of a gray-black desert jackal of a strange color, of white color."

This anecdote exemplifies one basic way of reaching understanding: matching a stimulus to a scheme or to soundly based knowledge. In the present case the scheme is visual and is located in Umlek's cognitive system but not in Bugamba's. Umlek understands that in the sky a picture of a bear appeared; he is fully aware that in the heavens there appeared the terrifying picture of the predatory bear; here a match is created (and for now I don't go into the question of how this happened) between a shape in the sky and the scheme of the polar bear in Umlek's head. Contrary to him, Bugamba doesn't understand what his friend is talking about, because the scheme of the polar bear has not yet been created in his cognitive system. In other words, in the present case the two are talking different languages. Each understands his own language but not the other's.

The Ability to Understand the Other from the Other's Viewpoint

Let us study the following fictional episode: three friends go to a party. Two of them (Dan and David) are standing at a soft drinks counter and looking at the partygoers. The third, Ron, goes to their left toward the toilet. Facing Dan and David stands a tall woman wearing a broad-brimmed hat. Dan whispers to David that but for the terrible scar on the left side of her face, which makes her look hideous, she would be regarded as a breathtaking beauty. After a while Ron returns and asks excitedly where the woman with the broad-brimmed hat is, because he feels he has fallen in love with her. Now David whispers into Dan's ear, "Tell me, has Ron gone mad? He goes and falls in love with that repellant woman?" And Dan answers in a whisper, "No, he hasn't gone mad. He saw her from the right side, her good side. In his place I would have fallen for her too."

This tale exemplifies another important way of reaching understanding: perceiving the other's behavior not from the viewpoint of the observer himself but from the other's, the one being observed. In the present case, misunderstanding how Ron has fallen for a repulsive-looking woman is made clear by the transition from the viewpoint of the two friends, Dan and David, to Ron's viewpoint—from his viewpoint it was not possible to see the dreadful scar on the woman's face, and the profile that Ron saw was that of a beautiful woman.

In this example I have used in the simplest way a process called empathy, that is, the ability of one person to put himself in the shoes of another, to understand him from his viewpoint. I do not intend here to go into a discussion of the modern development surrounding this subject in the two theories Theory-Theory (TT) and Mental-Simulation Theory (ST), or into the empirical/theoretical debate that took place between them, because such a discussion exceeds the purpose of the present chapter. Very briefly, according to TT, people have a popular theory that also contains mental concepts such as will and belief, by means of which they explain the other's behavior. According to ST, people represent the mental processes of the other who is caught up in a certain situation by means of cognitive simulation, namely by internal operation of similar mental processes in their own cognitive system. (On this, see a review and discussion in Gordon, 2009.)

Understanding through Sense Realizations

Visual realization of free fall. In his 2009 article, Lipton describes a thought experiment of Galileo as an example of the ability to reach understanding without resorting to a scientific explanation (but see the debate on this matter in Khalifa, 2013, and Strevens, 2013). According to Aristotle, a heavy stone, S_H, will free fall faster than a light stone, S_L. We tie the two stones together with a rope and drop them as a single body. If Aristotle is right, the body ($S_H + S_L$) will fall faster than S_H, because the body is heavier than S_H. However, because the light stone falls slowly, it will slow the fall speed of the heavy stone, and therefore the body ($S_H + S_L$) will fall slower than S_H! A logical contradiction is created here (the body will fall faster but also slower than S_H), and this leads us to the understanding that weight has no effect on the speed of free fall. Can this understanding be demonstrated? In my opinion the answer is affirmative.

In the first stage of the demonstration, two objects are dropped from the same height: a small iron ball and a sheet of paper, where the latter is evidently much lighter than the iron ball. In this demonstration it is seen at once that the iron ball hits the ground faster than the sheet of paper. In the second stage, one first takes the sheet of paper and folds it over and over until it forms a small, symmetrical ball the size of the iron ball. (Here I don't get into the question of how this is done in practice.) Again, the paper ball is far lighter than the iron ball. Then the two balls are dropped from the same height. This time it is clearly seen that the two balls hit the ground at the same time.

This realization quite naturally leads the observer to three conclusions. First, it is understood that the air slowed the fall of the sheet of paper more than the fall of the iron ball, because the area of the sheet's contact with the air was larger than the area of the iron ball's contact with it. Secondly, when the area of the sheet's contact with the air was reduced to equal the area of the iron ball, both balls fell at the same speed. And thirdly, because the two balls dropped at the same speed despite their difference in weight, it arises that weight has no effect on the speed of free fall.

Realizations through experiments in psychology. One of the experiments that made a great impression on me in the first year of my studies in psychology at the Hebrew University in Jerusalem was Masserman's (1946) experiment to create experimental neurosis. In the first stage of the experiment, a cat learned to press on an electric switch that opened the lid of a food box in which the hungry cat found food—a positive reinforcement. In the second stage, after this learning had become well-founded, when the cat put its head into the food box, it unexpectedly got an air blast in its face. As a result, the cat's behavior changed in an extreme way: even after two days of starvation, the cat refused to touch the switch that opened the lid of the food box.

This realization aims to create in the laboratory behavior parallel to neurotic behavior. On the one hand, this is a phenomenon that itself needs an explanation: Should the cat's behavior in preferring to suffer hunger rather than endure a sudden blast of air be seen as a kind of measure of the fear acquired by the cat in Masserman's experiment? (Rakover, 1975, showed in a psycho-physical experiment similar to the one described here that the level of conditioned fear was equal to one-third of the intensity of the electric shock that conditioned this fear in a rat.) Is a conflict created here of the approach–avoidance kind? On the other hand, to understand what happened, the researcher must take into account these questions and clues in order to propose a theory that provides understanding. Actually, almost all experiments that psychologists conduct (e.g., in classical conditioning, instrumental learning, visual perception, social psychology, etc.) exhibited two properties: the experiments create phenomena that require understanding, where these phenomena contain certain clues to setting out on the path leading to understanding.

Causal Factors

Visual illustration of the episode of the slanting pole. Understanding *why* and *how* questions sometimes results from observation of a situation of the kind shown in Figure 1.1: an iron pole, a mast, slanting sideways and held by a rope. This episode of the slanting pole may give rise to three possible causal interpretations, depending on the way we perceive and conceptualize

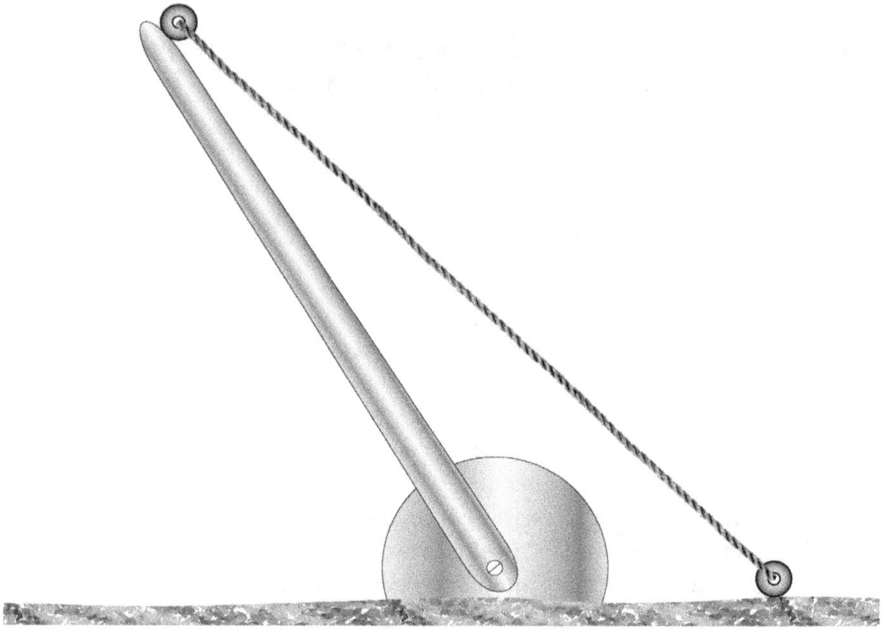

Figure 1.1 The Slanting Iron-Pole.

the factors responsible for the iron pole not falling to the ground. By the first interpretation, what meets the eye in Figure 1.1 is this: when the rope is untied, the mast will fall. The question arising is what will cause the disconnection itself. And here there are manifold possibilities: episode 1, someone cuts the rope with a knife; episode 2, a strong wind, a storm, hurls a chunk of broken wood that breaks the rope; episode 3, the iron pole sustains a powerful blow to its top (some heavy object crashes down onto it), severing the rope. Each of these three possibilities, from three episodes, comprises a number of events that together cause the separation of the rope and as a result the fall of the iron pole to the ground. We may denote these possibilities in an abstract form:

1. $[a \,\&\, b \,\&\, c \,\&\, \alpha] \rightarrow P$, when a, b, c represent factors relevant to episode 1, α represents the rope disconnection, P represents the iron pole falling down, and $\rightarrow$ represents causal relation.
2. $[d \,\&\, e \,\&\, f \,\&\, \alpha] \rightarrow P$, when d, e, f represent factors relevant to episode 2.
3. $[g \,\&\, h \,\&\, i \,\&\, \alpha] \rightarrow P$, when g, h, i represent factors relevant to episode 3.

If we examine these three episodes by the *method of agreement* proposed by Mill (see discussion in Copi, 1968), it arises that α is the factor common to the three episodes, and therefore it will be considered to be the cause of *P*. But this conclusion seems hasty, because of the following interpretation.

It may be suggested that α is a kind of Insufficient, Nonredundant, Unnecessary, Sufficient (INUS) condition, in Mackie's (1974) view. By his view, the breakage of rope α is not a sufficient condition for the fall of pole *P*, because for the rope to be detached several events must take place that together constitute such a sufficient condition. α then is a Nonredundant condition in the group of factors of episode 1. However, Mackie then states that this group is an unnecessary condition for the pole to fall, because there are other groups of factors that entail this fall. Quite clearly, what causes the fall of the pole is gravity. Without it, the disconnection of the rope would have no effect. In other words, α allows gravity to operate, because until the disconnection, the rope acted as a counterforce to gravity. But in addition to this, we must also take into account the following possibilities: the fall of the pole will not be prevented by a support beneath the pole or by operating a highly powerful magnet above the pole. In brief, α is one factor in a group of additional factors (gravity, absence of support, and nonoperation of a magnet), all of which together constitute a sufficient condition for *P*.

Realization by perception of causality: In 1946 the Belgian psychologist Michotte published a book summarizing his research on the "perception of causality"—famous experiments that became classics (in 1944 Heider and Simmel published an article on the same subject; see discussion and review in Rakover, 1990; Scholl & Tremoulet, 2000). Contrary to the perception of Hume, who interpreted a causal relation as a process of learned inference, Michotte suggested that perception of causality is a process of innate inference. As an example, we shall look at the following simple illustration: circle A moves from left to right on a horizontal line, reaches circle B, touches it, stops, and circle B begins to move from left to right on the horizontal line. Participants in this illustrative experiment described what their eyes saw in causal terms: A pushed B, which as a result began to move in the same direction previously taken by A.

In a conclusion arising from their extensive review of dozens of experiments, interpretations, and discussions, conducted following the initial studies of Michotte and of Heider and Simmel on the phenomenon of "perceptual causality and animacy," Scholl and Tremoulet wrote the following (countering the critique of Michotte's conclusions): "These interpretations seem to be perceptual in nature—to be fairly fast, automatic, irresistible and highly stimulus driven—despite the fact that they involve impressions typically associated with higher-level cognitive processing" (2000, p. 299). And also "such phenomena reflect primarily perceptual and perhaps module processing" (p. 307).

Realizations That Intensify the Use of Simple Inferences to Reach Understanding

Arithmetical queues: Assume that the participant in a realization is presented with four boxes, in two of which are balls; in the fourth box is a question mark:

[ ] [**] [****] [?],

Where ? signifies the question: How many balls are in this box?
"The answer is very simple," the participant says: "Six." "Correct," replies the experimenter.
And now appears a similar question, when only three boxes are presented:

[**] [****] [?]

This time, too, the participant replies, "The answer is very simple: six." But this time the experimenter replies, "Correct, the answer could be six, but also eight." "Aha," says the participant, "in the first queue the rule was this: the number of balls in the next box is equal to the number of balls in the previous box *plus* two. The rule in the second queue could be the rule for the

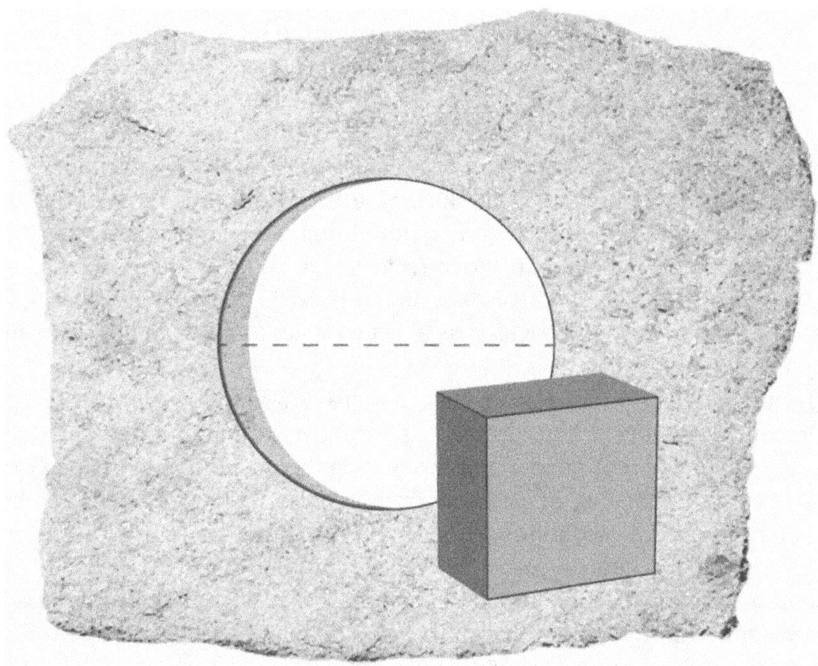

Figure 1.2 The Circular Hole and the Square.

first queue, but also a new rule: the number of balls in the next box is equal to the number of balls in the previous box *times* two." "Very good," says the experimenter.

The square and the circular hole: Figure 1.2 shows an inflexible block with a round hole in it through which one wishes to pass a square body of rigid material. The question for understanding is this: When will the square pass through the hole, and when will it not? The answer, as can be seen at once from the figure, is that the square will pass through the round hole when its diagonal is smaller than the circle's diameter. (This example follows the "cubical peg, the square and the round holes" example. See Putnam, 1975b, chapter 14.)

A provisional conclusion: Considering the above examples, the following question arises: What is the structure of the procedure, the model, for understanding common to all these illustrations? My proposal, which I shall call the General Understanding Procedure (GUP), is that the GUP model (scheme) is based on two basic components:

1. It is assumed that every phenomenon that requires understanding has some unknown process responsible for its occurrence (an Unknown Real-Process [URP]);
2. It is assumed that an oft-repeated examination and illustration of the given phenomenon in different ways will lead to its understanding, in that the phenomenon will become cognitively bright (e.g., it will become clear that the phenomenon is not the outcome of chance), and its components, factors, and conditions involved in its occurrence will be highlighted. As a result, the extent of understanding the phenomenon (i.e., the answers to the *why* and *how* questions) will increase.

And now in light of the GUP model, we shall examine the examples I discussed above.

Both Umlek of the North Pole and Bugamba of Africa try to understand the shape of the clouds in the sky. They assume covertly that something has caused the clouds to take on that shape. Umlek is convinced that this shape is closely bound up with the polar bear, because it is very similar to the shape of the polar bear that savaged two children of his tribe. But while this understanding speaks from Umlek's deepest feelings, it is silent in Bugamba, because the scheme of a polar bear has not been created in his head.

Neither Dan nor David understand how Ron could possibly have fallen in love with that hideous-looking woman. Something caused it, but what? The solution lies in an attempt to see the woman from Ron's viewpoint—this is a very simple and clear solution: from that viewpoint Ron could not have seen the awful scar on the left side of the face of the woman with the broad-brimmed hat.

Aristotle held that heavy objects fall faster than light objects. Was he right? In this case too, doubts presumably arise in connection with URP. This question is tackled by means of a realization (experiment) in two stages: first, free fall of a small iron ball and a sheet of paper, and then of two balls of the same size—one of iron, the other of paper. The results of this demonstration attest to Aristotle's mistake.

Masserman's experiment realizes empirically a phenomenon that may be interpreted as neurosis based on fear and the approach–avoidance conflict. Here too the question arises as to what the process responsible for creating this abnormal behavior is (what is the URP?). The answer to this understanding-question is very complex, and my impression still today is that there is no theoretical agreement on the subject (even though the literature is awash with hundreds of experiments, theoretical articles, and books). What is clear is that the element of the conditioning of fear is linked to the dramatic change in the cat's behavior.

The slanting pole episode demonstrates well that the fundamental question is connected to the unknown process responsible for the occurrence of the given phenomenon, so the question underlying this episode is this: What is likely to happen? The three possible episodes that propose various causal explanations as to what is likely to happen are different ways of reaching understanding: What is the factor responsible for the fall? What prevents the pole from falling to the ground? And how should one regard the different answers from the viewpoint of this so very complex concept—the concept of causality?

Michotte's series of experiments on the perception of causality asks a question different from the questions arising from the slanting pole episode. Here the basic question (the URP question) is: Are we inclined to take what is shown in Michotte's experiment as something we have learned every day of our lives? Or perhaps, because of the characteristics of the participants' responses in the experiment (e.g., fast reaction time, automaticity), must we assume that responsibility is borne by a modular cognitive process specializing in perception of the relations between stimuli in causal terms?

Realizations related to guessing the next item in the queue presented to the participant are closely connected to the URP question: How is the queue constructed? What is the rule whereby the queue is constructed? The instant the answer is found, understanding will be induced within us.

Something similar happens with the illustration of the square and the circle. The question is under what condition will the square pass/not pass through the hole in the circle? The answer that the square will pass through when its diagonal is smaller than the circle's diameter induces understanding in us. (This problem is very similar to the road sign "A vehicle higher than 2.5 meters may not enter the tunnel.")

Relations between Procedures of Understanding and of Explanation: Differences and Cooperation

Differences

What are the basic differences between procedures that provide us with understanding and those that provide us with explanation? The following chapters survey the theories, models, schemes, and various procedures for presenting explanation that have inundated the professional literature since the publication of Hempel and Oppenheim's (1948) Deductive-Nomological (D-N) model. Yet already here the question before us can be answered if we take into account the following two matters: first, in practice all empirical researchers in psychology (and in the other sciences) have learned how to propose explanations; furthermore, in their research they have also proposed explanations for the given phenomena, so presumably they know what presenting explanations in science involves. Secondly, to clarify here formally what is under discussion, I shall summarize very briefly the explanatory D-N model: this model posits that if, from a certain law (a given theory), together with data of a certain situation (in psychology, a description of the independent variable), one infers a prediction that matches the observation under study (in psychology, a description of the dependent variable), we may say that the observation has obtained an explanation by this law (this theory). Considering the above, I propose the following basic distinction: *Contrary to an understanding-procedure, an explanation-procedure is interwoven with empirical scientific methodology, developed over the past five centuries, and constitutes an integral part of it.*

This statement carries several important implications that I consider here.

First, understanding-procedures attempt to induce understanding of the given phenomenon (i.e., its representation in language) by connecting the phenomenon to *deep-rooted knowledge, which is accepted unquestioningly* (e.g., knowledge about the world founded on religious beliefs, social rules, and political regime). But the explanation-procedures aim to enhance clarification of the studied phenomenon by connecting it *to knowledge that has withstood the test of scientific methodology* (which has been developed and perfected for the past several centuries). I shall call the connection between knowledge that enhances understanding of the given phenomenon and the phenomenon itself "the understanding–phenomenon connection," and the connection between knowledge that enhances explanation of the phenomenon and the phenomenon itself "the explainer–explained connection."

Second, the understanding–phenomenon connection is realized through demonstrations (mainly visual), through drawing lines of similarity between the phenomenon and deep-rooted knowledge, by use of analogies and metaphors, and sometimes by acceptance of an authoritative opinion without demur (such as religious authority, ruling authority). By contrast, the explainer–explained connection is realized through use of three principal methodological means that are subject to criticism by the scientific community:

1. *Rationality*: The explainer–explained connection, the connection between relevant knowledge (the information that explains the given phenomenon) and the given phenomenon, rests on a rational foundation. That is, the explainer–explained connection is anchored to logic, mathematics, the law of probability, a theory of decision-making, induction, practical considerations, and the like.
2. *Empirical test*: Contrary to the understanding–phenomenon connection, the explainer–explained connection must pass a systematic empirical test. The absence of a demand for an empirical test in the understanding-procedure is liable to create a wholly wrong understanding. For example, although some stars look like a bow or a great bear, and the sun looks as if it is traveling over the sky around the earth (a perception that suited the doctrine of the Catholic Church), we know today that it isn't so.
3. *Measurement procedures*: There is a great difference between the natural sciences (physics) and the social sciences (psychology) in regard to the explanation/understanding and the studied phenomenon: measuring units are still not discovered in psychology as in physics (see Rakover, 2012b). This difference becomes greater by far when it is applied to the distinction between the understanding–phenomenon connection and the explainer–explained connection. To understand this, I now move on to a short discussion highlighting the basic problem of measurement in psychology.

An important difference between psychology and physics resides in the attempts to connect the concepts of the theory to the concepts of the observation, the phenomenon to be explained (call it the "theory–observation gap"). The difference lies in the following measurement property. While physics bridges the theory–observation gap by using *real* measurement units (e.g., a ruler to measure distance and a balance to measure weight), psychology bridges the gap by using *hypothetical* measurement units that are indexed by one's responses. What is measurement in physics? Here I shall describe measurement as the relation between a certain quantitative property of a given object and a unit of measurement of this property, which is revealed empirically (e.g., Coombs, Dawes, & Tversky, 1970; Michell, 1999). For example, if we have a straight stick of length A, and we find that another stick of length M (which is arbitrarily determined as our unit of measurement, e.g., the meter) goes into A ten times from end to end, then it is discovered empirically that length A is ten lengths M (i.e., $A/M = 10$). Given this, the main point I want to emphasize is that scientists have found an empirical operation (counting how many times *M* goes into the length of the measured object), which upholds mathematical properties that define the mathematical language on which the physical theories are based. In view of this, it emerges that what one says by means of mathematics will be said exactly by means of using the real measurement units. And this, I believe, is the strongest tie that can be made between theory and observation. (Similar things can be said about several other quantitative properties, such as weight and time.)

Can one suggest that in psychology the connection between theory and observation is based on the measurement procedure described above? My answer is no. The main reason for this negative answer is as follows: no real measurement units have been discovered in psychology, since all kinds of behavior, which involve one's goals, intentions, free will, and the like, are no more than certain behavioral expressions, indexes, of one's inner mental, cognitive, processing (conscious and unconscious). How can psychology know about these inner processes? The answer is that the science of psychology can know about them only by generating hypotheses and theories. That is, there are no real measurement units for our inner life (whereas in physics the theories are based on certain concepts that translate directly into real measurements). For example, for our conscious experience no unit of measurement has as yet been found as it has for length: it is not possible to define a real measurement unit of love (the meter of love, M_{love}), and say that Jacob loves Rachel ten and half units of love more than he loves Leah. By this example I do not mean to suggest that one cannot say that Jacob loves Rachel much more than Leah, only that a real measurement unit of love (M_{love}) has not been discovered like the natural unit of length (the meter), and therefore one cannot say, for example, that Jacob's love for Leah is $10M_{love}$ while his love for Rachel is $20.5M_{love}$ (for further details, see Rakover, 2012b).

Cooperation

Despite these major differences between understanding-procedures and explanation-procedures, it merits stating that the scientific methodology practiced in the sciences, more particularly in psychology, uses several of the understanding-procedures as important aids that provide clues to the construction of a good scientific explanation for the given phenomenon. For example, one may take the illustration of free fall to exemplify the independence of the weight of the falling body from the speed of the fall, where the explanation for this phenomenon is through an appeal to the Newtonian theory of gravity. In psychology, Masserman's experiment can be used, on the one hand, as an exemplification of the creation of fear and conflict, and, on the other, as an experimental template that facilitates different changes and variations that serve to affirm or reject diverse hypotheses whose purpose is to explain the cat's strange (neurotic) behavior. And another example: it turns out that even behaviorist psychologists (e.g., Tolman, Hull) used the empathy process to suggest explanations for the behavior of rats in their experiments (see Eigner, 2009).

UNDERSTANDING, EXPLANATION, AND MODELING

Modern science used models of different kinds (some of which applied visual illustrations) to facilitate the formulation of theories intending to explain the studied phenomena. For example, Morrison (2009) suggested that models in science (she referred mainly to mathematical models in physics and biology) deepen the understanding of the relation between the theory and the phenomenon to which the theory is applied. A model offers a simplification that allows application of the theory, and also abstract representation, idealization, of the given phenomenon. Furthermore, Lenhard (2009) proposed using computer simulations (a model operable by a computer and imitating certain characteristics of the phenomenon) to enhance understanding by systematic activation of simulation, which makes it possible to observe and learn how the computer model functions. This approach, which emphasizes the importance of the model in an attempt to understand the studied phenomenon, finds expression in what de Regt (2009) writes: "Scientific understanding of phenomena (UP) requires intelligible theories, where intelligibility is defined as positive value that scientists attribute to the theoretical virtues that facilitate the construction of models of the phenomena" (p. 37).

An Example: The Catch Model for Face Reconstruction

I conclude this chapter with a discussion of an empirical/theoretical illustration, the Catch model, with the aim of demonstrating the entire issue of the relation of theory, model, and phenomenon to understanding and explanation. In developing this model, we strove to understand how it is possible to reconstruct from the memory of an eyewitness a face he has seen previously (a suspect's face, a target face) (see Rakover & Cahlon, 1989, 1999, 2001, 2008). The basic idea behind this illustration is as follows.

The Catch model may be analyzed as based on two submodels. The first has to do with the given phenomenon, and I denote it "M-phenomenon" (model of a phenomenon). It represents only certain aspects (properties, processes) of the phenomenon and overlooks others. This submodel becomes the event that one is trying to explain. For example, in Masserman's experiment no effort is made to explain the entire complex of the cat's behavior, but basically if the cat opened the lid of the box in order to obtain a positive reinforcement.

The second submodel, which I denote "M-explanation" (model for explaining M-phenomenon), aims to offer an explanation for M-phenomenon. In this sense, the overall model, the Total Model, the Catch model, may be regarded as based on M-phenomenon and on M-explanation. Here, the two following comments are apt:

1. The explanation of Total Model is empirically testable. That is, M-phenomenon is nothing other than a partial and selective representation of the phenomenon. It represents certain aspects of the phenomenon itself, representations linked to the observations by operational definitions or by a measuring process. Therefore, the prediction emanating from M-explanation about the behavior of M-phenomenon is testable against the observations.
2. The Total Model can be perceived as a theory in the framework of the semantic approach to a scientific theory (see Rakover, 1990). Furthermore, in psychology, models are grasped as small theories, because in most cases in psychology there is no broad theory that fosters perceiving a model as a realization and a simplification (idealization) of a theory aimed to facilitate the treatment of the given phenomenon. For example, the Catch model is not a model of a broad theory for perceiving faces (to the best of my knowledge no such theory has as yet been developed) but is itself a kind of limited theory to reconstruct faces from the witness's memory, where this model is developed in the general conceptual framework of cognitive psychology: information processing (see Rakover & Cahlon, 2001).

The Catch Model

The motivation to develop the present model stemmed from the failure of police methods (e.g., the Identikit and the photo-fit) to reconstruct from an eyewitness's memory the suspect's face (the target face). By these methods an attempt is made to assemble the target face by choosing out of many dozens of facial features (e.g., eyes) the one most similar to the face the witness has seen before. Thus, one constructs from the most similar hair, forehead, eyes, nose, and chin the face that is taken to be most similar to the target face. The problem is, as I mentioned above, that these techniques have not proven particularly successful. The reconstructed face is not like the target face. One of the reasons for this failure, we thought, is connected to empirical–theoretical findings that people recall faces not as a collection of facial features but as a single unit—that is, they remember a face as a whole. The Catch model is based on this notion: instead of building the face from parts, which the police methods do, we concluded that we needed to collect a large number of faces that seem to the witness similar to the target face, to analyze these faces for their components, and then by a suitable procedure (to be described later) to locate the face components belonging to the target face, and by their means to compose that face. Considering the foregoing, and in order to describe the Catch model in the clearest possible way, I shall first describe how we conceptualized the face mathematically, the structure of the experiment to test the model, its empirical/theoretical testing, and finally the implications arising from it.

Mathematical Conceptualization

We treated facial features such as hair, forehead, eyes, nose, and chin as facial dimensions, and we denoted the number of possible dimensions m (e.g., in the example at the beginning of this sentence $m = 5$). Every dimension has a large number of values (denoted n). For example, the eyes dimension may contain a large number of different shapes of eyes: round, elongated, downward slanting, upward slanting (like the eyes of Chinese people), etc. The number of faces that may be created from m dimensions and n values for each dimension is given by (face number) $Fn = n^{m}$. For example, where $m = 5$ and $n = 3$, we obtain $Fn = 3^5 = 243$. (Clearly, the number of faces increases very rapidly as n rises.)

Experiment Structure

The experiment is based on the formula of number of faces: $Fn = n^{m}$. For example, if $m = 5$ and $n = 3$, we obtain a pool of 243 faces. From these we shall choose one face, which will serve as the target face. We shall thus be left with 242 different faces, by means of which we will reconstruct the target face from the witness's memory. The experiment itself has two stages. In the first, the target face is exposed for several seconds. In the second stage the participants in the experiment are presented with a series of pairs of test faces sampled from the pool Fn. The number pairs (Pn) that can be created from Fn is given by the formula $Pn = [1/2(Fn - 1)(Fn - 2)]$ (as can be seen, in calculating the number of pairs we subtracted the target face). Because the participants cannot be presented with all the pairs, they are shown a random sample of them. Each participant's task is to choose from each pair the face most similar to the target face presented to her at the start of the experiment.

The Catch Model: Basic Assumptions

The model rested on two basic assumptions: (a) the choice rule—the way the participant chooses the test face most similar to the target face; and (b) the reconstruction rule—the way one should analyze the chosen test faces as most similar to the target face in order to reconstruct this face (for further details of this model, see Rakover & Cahlon, 1989, 1999, 2001, 2008). (For illustration, see Figure 1.3.)

The choice rule: Assume that the number of facial dimensions is $m = 5$ (A hair and forehead, B eyes, C nose, D mouth, and E chin) and $n = 2$, and assume that the target face is $A_1\ B_1\ C_1\ D_1\ E_1$. For simplicity, we shall denote this face

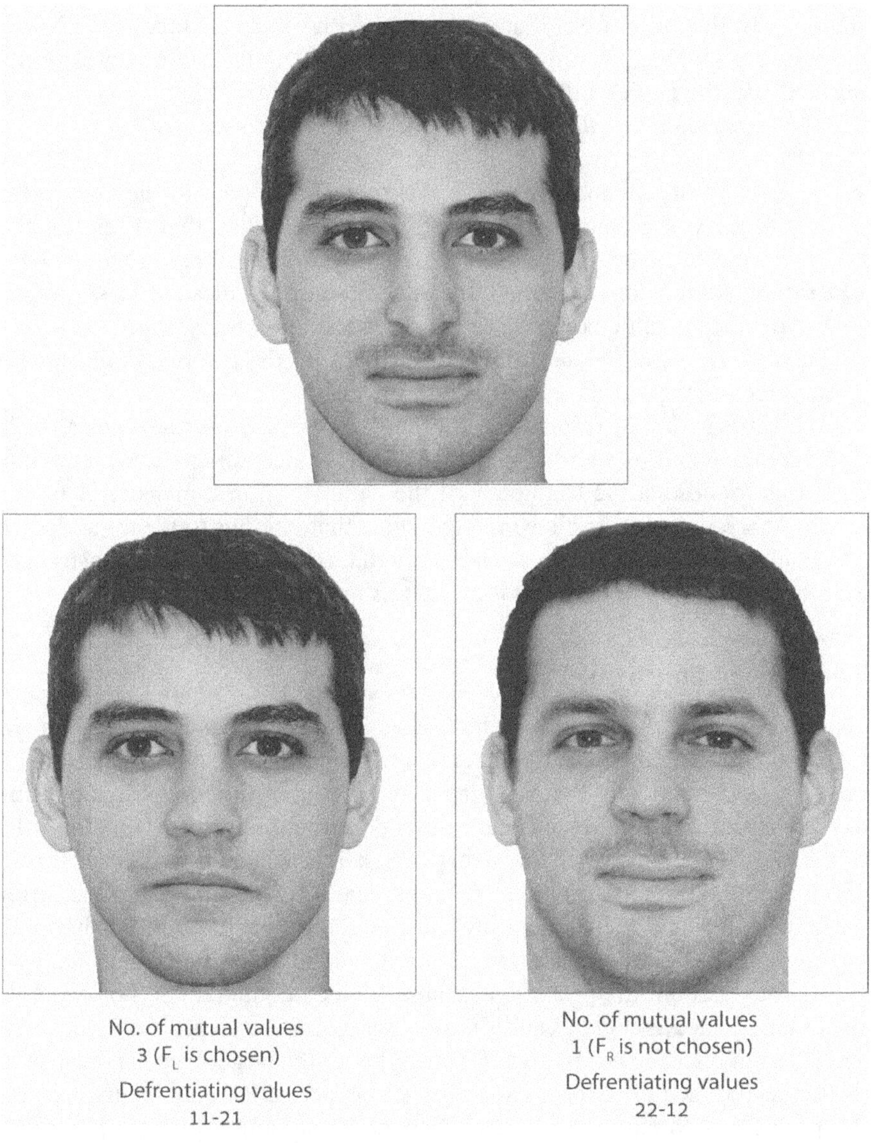

Figure 1.3 A Target Face with a Pair of Test Faces.

11111. In keeping with the experiment structure, the participant is shown a pair of test faces, for example, on the right, face F_r: 22212, and on the left, face F_l: 11221 (see Figure 1.3). The model assumes that the participant has perfect memory (i.e., the participant recognizes the target face flawlessly) and there is no cognitive or other interference in performing the following actions of the choice rule: the participant compares each test face with the target face and counts how many facial features there are in every test face that are shared with the target face, that are mutual with the target face, and chooses the test face that has the largest number of such shared facial features (in the present case the choice falls on the left-hand face, F_l: 11221).

The reconstruction rule: The rule is based on three sections:

1. For every test pair the values are recorded that appear in the chosen test face (e.g., in the example above, F_l was chosen) and that do not appear in the rejected face. These values are called "differentiating values." For example, in the above example the differentiating values are 11–21, where "–" denotes a value common to the right and the left test face.
2. The differentiating values are summed over all the test pairs presented in the second stage of the experiment.
3. The differentiating values of the highest frequency are the values of the target face, and by their assembly this face is reconstructed. For example, if it is found that the frequency of the values 1, 2 in dimension A is A_1 = 73, A_2 = 35, then A_1 is chosen as the value that belongs to the target face in facial dimension A. (Note already here that section 3 in the reconstruction rule requires theoretical proof as well as empirical testing.)

Theoretical–Empirical Test

As noted in parentheses above, the test of the model was conducted on two levels: theoretical and empirical. As regards the theoretical, we succeeded in proving mathematically that with the assumptions of the model it is possible to reconstruct the target face. That is, it is possible to discover what the value of the target face is for each and every dimension, when all possible test pairs (excluding the target face) are tested, and when it is assumed that the different dimensions and their values are independent of each other. Furthermore, we were able to prove that it is possible to reconstruct the target face even taking into account the problematic pairs, called "equal pairs." For example, if the target face is 11111, and the two test pairs are right: 11233 and left: 23211, each test face has two values shared with the target face, so one cannot decide which face to choose. Pairs of this kind are called equal pairs. Clearly, the problem with equal pairs stems from the assumption that the perceptual weight (the influence on perception of the face) of all the dimensions is identical. For example, if the test face has a shared value with the target face in the chin dimension, the weight of this value (one point) is identical to the weight of the shared value in the eyes dimension (also one point). However, it was possible to improve the explanatory power of the model if the dimensions were weighted and graded (e.g., hair and forehead > eyes > nose > mouth > chin, where ">" signifies "has a greater weight than"; a mathematical proof of this appears in Rakover & Cahlon, 1999, 2001).

The results of several empirical experiments conducted in the laboratory supported the model very well: the target face was reconstructed in full or very closely, when the number of facial values was small (n = 2, 3). The empirical success of the model declined very quickly when n increased. This was a serious problem (we call it "the big n problem"), and a variety of efforts to resolve and explain it failed. These included an interesting attempt to develop four more rules for reconstructing the target face. Although these four led theoretically to complete reconstruction of the target face, they could not solve the present problem (see a detailed account of these issues in Rakover & Cahlon, 2001). When these exertions ended in great disappointment, and when our repeated checking of the findings ended, an interesting supposition arose: the above assumption (no cognitive interference or other interference in performing the choice rule) was apparently mistaken. In consequence, a new assumption emerged, that the multiplicity of test faces and their parts, due to the enlargement of n, created a retroactive interference to the remembering of the target face, and as a result the success of the reconstruction steadily declined (see Rakover & Cahlon, 2001). To overcome this interference, it was decided to check how far recognition of similarity groups consisting of facial features like those of the target face was likely to help in reconstructing a small group of faces similar to the target face, including the target face itself. (In practical terms, this may greatly assist police investigators in their work.) As an example of such a similarity group we shall examine the "hair and forehead" facial dimension.

Assume an experiment based on the fact that every facial feature contains three similarity groups, where in each group there are three values. Let us look at the "hair and forehead" dimension. Group 1, "thinning hair," has three values: baldness, very thin hair, thin hair; group 2, "short hair": very short hair, medium-short hair, short hair; group 3, "long hair": hair to the shoulders, hair to the chin, hair to below the ears. Assume also that the target face is bald and therefore "thinning hair" is the target similarity group. And now, after the participants have undergone the above procedure—first they saw the (bald) target face and responded to the test face pairs—the experiment moves on to the stage of analysis of participants' responses as described by the *reconstruction rule*. But now, in addition to the identification of the facial features of the target face, one more possibility is added for identifying the similarity group of the target face—the group in which the facial feature of the target face appears. In the present example, because the target face is bald, its group is thinning hair: baldness, very thin hair, thin hair. That is, even if the analysis of the results does not produce the value "baldness" but produces the value "very thin hair," we may say that we have succeeded in locating the similarity group of the target face for the dimension "hair and forehead." (If, for example, the analysis had produced the value "very short hair," it would mean that for the dimension of the present face, the analysis did not succeed in identifying the value of the target face and the similarity group of the target face.) The results of a series of experiments with similarity groups went against the following intuition: the likelihood of reconstruction of the similarity groups of the target face will be greater than the likelihood of reconstruction of the values of the target face itself. However, these results are readily explained by calculation of the amount of retroactive interference: the interference is greater in the condition of the reconstruction of the target similarity group than in the condition of the reconstruction of the target face (for details, see Rakover & Cahlon, 2001, 2008).

Theoretical Implications

The Catch model has two important implications: one concerns the reconstruction of information from memory, and the other concerns the connection of a scientific model to understanding and explanation.

Reconstruction of Information from Memory

In consequence of the development of the Catch model, Rakover (2005) proposed a deductive method for reconstructing information from memory, the Deductive-Reconstruction (D-R) method. The idea behind this method is based on two fundamental conditions:

1. *The Provability condition by the model*: It has to be proven mathematically that from (a) the *memory task* to be performed (e.g., choice of which of the test faces is most similar to the target face), (b) the *performance rule* for executing the memory task (e.g., a rule that determines how to choose the most similar face to the target face), and (c) the *reconstruction rule* that analyzes the outputs (responses) that the performance rule produced, it is possible to fully restore the target information presented in the past (e.g., a target face).
2. *The Matching condition by the participant*: If the participant's responses in the *memory task* are identical to the outputs (responses) that emerge from the *performance rule* applied in this task, it will be possible to

reconstruct the information from the participant's memory exactly as it was possible to reconstruct the information from the model's memory according to (1). In other words, reconstruction from the participant's memory is complete because it is based on analysis of her responses to the *memory task* (which are identical to the responses produced by the *performance rule*) and because in (1) it was proven mathematically that the reconstruction was perfect.

Note that these are the two fundamental conditions on which the D-R method is based. But in addition, the present method discusses the practical case where the degree of match of the participant's responses to the responses produced by the performance rule in the memory task (i.e., the responses that the model produces) is not perfect, and contains a number of errors. The principal question that this method considers (I shall not describe this consideration here) is this: What is the number of errors that may be made for it to still be possible to reconstruct the information from the participant's memory? (See Rakover, 2005.)

Models, Understanding, and Explanation

As may be gathered from the present chapter, the understanding concept is broader than the explanation concept. Although it is possible to accept the argument that understanding a phenomenon (Why did it occur? How did it occur?) depends on giving an explanation in line with the game rules of scientific methodology, understanding the explanation itself depends on understanding the explanatory theory or a model of the explanation, and on the connections of the explanatory theory to relevant theoretical/empirical methodological knowledge. As stated earlier, scientific models provide understanding and explanation of the studied phenomenon because they simplify the relevant information, and concentrate theoretical/empirical understanding on those aspects of the studied phenomenon that are assessed as most vital. For example, when we wish to know if an iron ball falling from a certain height will fracture a wooden surface below, we concentrate principally on the ball's weight, not its color (black, red, yellow).

The Catch model described above indeed simplifies the process of reconstruction of the picture of the target face and offers a way to understand the success/failure of the reconstruction. In addition to the model's concentrating attention on the most relevant properties for reaching the goal of reconstruction while overlooking other properties, it also makes several unreal assumptions. Although the latter distance the model from faithfully representing reality (they may be seen as distorting it), the model succeeds in advancing our scientific understanding of the subject at stake. (For the argument that scientific models aid in advancing scientific knowledge even when they are false, that is, models that are fictional, idealizations, but supply good explanations, see Bokulich, 2011.) To justify this last statement, I analyzed the Catch model as an overall model (Total Model) based on two submodels: the submodel of the phenomenon (M-phenomenon) and the submodel intended to explain the phenomenon (M-explanation).

The M-phenomenon centers on the choice response between two test faces according to the criterion of similarity to the target face. The focus on this response disregards the other stimuli, responses, and processes that are linked to the participant's internal cognitive system, and are liable to affect his behavior one way or another. (The Signal Detection Theory [SDT] attempts to deal with some of these factors, such as the individual's motivation and assessments of the situation in which he is. The SDT measures deal with effects of this kind, and they of course improve the situation on a certain level, but as may be understood, these measures too will be hard-pressed to deal with the negative effect of the retroactive interference on memory.) In short, M-phenomenon is a model constructed in the behaviorist spirit, which concentrates on the choice response itself, while disregarding the individual's intentions, the particular consciousness situation in which the participant finds himself or herself.

The M-explanation also is a kind of submodel that overlooks many factors, and makes several assumptions that intuitively are incorrect (they are far from reality). For the choice rule to function well, the M-explanation was based on several assumptions, such as perfect memory; absence of interference from the cognitive and other systems (e.g., emotionality) at the time of making the choice; independence of the face dimensions and their values; and an attempt to reconstruct the target face—in fact one of countless faces—with the aid of a small limited sample of faces. As for some of these assumptions, we succeeded with further developments of the model to approximate it to reality. For example, an attempt was made to evaluate the model's ability to reconstruct the target face by assuming an imperfect memory expressing deviations from the choices dictated by the choice rule (the question was how many mistakes can be made for it to still be possible to reconstruct the target face); similarly, it became clear that if weighting, grading, is applied to the face dimensions (hair and forehead, eyes, nose, mouth, chin) the efficiency of the model in reconstructing the target face improves. We did not deal with the other assumptions, and they continued to serve as a basis for the Catch model. Let us look at some such assumptions.

The assumption of independence of the dimensions was essential to make possible a mathematical proof, where given the model's conditions the face would indeed be reconstructed. In addition, the mathematical proof used all the possible pairs for m and n, when in fact in the experiments themselves the number of pairs was limited. The assumption of no interference turned out to be extremely dangerous: retroactive interference arising from the increase of *n* lowered the efficiency of the reconstruction very quickly. These things elicit critical questions as to the methodological status of the Catch model. For example, is this model able to explain the experimental findings? And how is it possible that a model based on ideal assumptions far from reality, that is, a model that can be deemed false, can succeed in giving an explanation for the empirical results? These are not specific questions arising only in connection with the present model, but questions asked with regard to any scientific model. For example, Bokulich (2011) argued that although Bohr's famous hydrogen atom model is assessed as an incorrect, fictional model (it does not accord with the picture that emerges from quantum theory), this model provides a good explanation regarding the properties of hydrogen gas. My answers to these questions in connection with the Catch model are the following:

First, the Catch model is not ad hoc. That is, it is not a model made to suit known results, and for a very simple reason: the empirical findings reported in the literature by Rakover and Cahlon (1989, 2008) were obtained after the Catch model was developed—that is, those findings were not published earlier; Rakover and Cahlon's publication was primary. This, then, is an explanatory model proposing clear and good predictions as to what is expected to occur given certain conditions (when *n* is small).

Secondly, the very success of the model in actual reconstruction of the target face (for small *n*) supports the hypothesis that the choice rule indeed closely describes an actual cognitive process. Moreover, the findings showed that in these cases there is a match: the participant's choice responses match the responses that emanate from the choice rule. Still, I do not sternly insist on this, because several other choice rules may be suggested that will also bring about target face reconstruction, and experiments have not yet been done that determine which of these rules best suits the way the participant's cognitive system operates (see discussion in Rakover & Cahlon, 1989, 2001). (It is worth noting that Rakover and Cahlon also described additional reconstruction rules.)

Thirdly, even though it became clear that large *n* reduces the model's ability to reconstruct the target face, continuation of the research induced understanding of the phenomenon. The hypothesis arose that the reason for the failure of reconstruction of the target face was retroactive interference acting on the memory of this face. The results of a number of experiments done in consequence of this hypothesis supported the development of a new model (see Rakover & Cahlon, 2001, 2008).

These comments lead to the following conclusion about the relation between understanding and explanation: explanation seeks by rational means to match the studied phenomenon to scientific knowledge that satisfies the methodological requirements of science; understanding comes about as a result of the explanation, but also of other factors involved in the research process itself, including successes but also failures. The attempts to understand the failures (e.g., of the Catch model in the present example) open new research paths that greatly advance scientific knowledge; that is, the attempt to understand the failure quickens understanding of new phenomena, which are revealed as research on the given subject continues to develop.

Chapter 2

Models of Explanation (I)

Natural Sciences

Before I embark on a discussion of theories, models, procedures, schemes of explanation, I must answer this natural question: What is a scientific explanation? The answer is complicated. On the one hand, scientists use different forms and kinds of explanation acquired while learning their particular profession and in their research work aimed at understanding the phenomena in question. On the other hand, philosophers of science analyze research work so as to understand the overall pattern of the supply of explanations. They also aim to suggest logical, epistemological, methodological criteria that they believe a scientific explanation should satisfy. So clearly, there is no general agreement as to the nature of a scientific explanation; and indeed, in the following pages it will be seen that different researchers propose different definitions for this complex notion of explanation. Still, so I believe, most researchers will agree that a scientific explanation intended to provide understanding must give answers to the questions *why* the given phenomenon has occurred and *how* it has occurred (*why/how* questions). That is, a scientific explanation may be seen as an answer to *why/how* questions, which are firmly anchored to scientific methodology with the aim of providing understanding.

This brief foreword raises a new question that I must deal with as well, before beginning the discussion of models of explanation: Why does this phenomenon—and not that one—create a demand in the researcher's heart for an answer to, an explanation for, the questions of why/how the phenomenon occurred. I hold that a phenomenon awakens attention when it constitutes a change relative to its condition previously or relative to the environment in which it is. For example, if David arrives at work every day at eight precisely, the question arises why precisely today he got there four hours late. If for years a house stood on a hill and today we discover that it isn't there, we shall ask why it has been demolished? If toward evening everyone in the neighborhood goes to the pub except Mr. Smith, the question arises why Mr. Smith is an exception—he doesn't go to the pub. The change in the situation arouses pondering as to why/how; and this pondering, where there is no answer in existing knowledge, sparks the wish for deep study that eventually will provide an answer to our questions, and the demand for understanding will fade away (see a similar idea in Lambert & Brittan, 1992; Salmon, 1990).

Philosophy of science has proposed a considerable number of models for answering *why/how* questions. This book is not intended to critically survey all the different approaches to explanation, because my concern centers on the question of understanding and explanation in psychology, not in physics, chemistry, or biology. Hence, I shall briefly treat several models of explanation that seem to me relevant to research in psychology, and that in my opinion merit psychologists taking the time to become acquainted with, simply because these models have opened my eyes to a better understanding of what goes on in psychological research, as in the natural sciences. These models (schemes, theories, procedures) of explanation do not determine the content of the actual specific explanation given to a certain phenomenon that raises *why/how* questions. For example, the Deductive-Nomological (D-N) model does not state the content and structure of Newtonian theory (presented at the end of the sixteenth century, many years before the publication of this explanatory model). In the Newtonian case, this model constitutes a scheme that lets us know how scientific explanations were given to certain physical phenomena.

I shall begin with a review of the D-N model proposed by Hempel and Oppenheim in 1948. This model was considered a cornerstone in philosophical research into explanation, and development in the field may be seen as implications emanating from it. In the discussion of this model, and of others, I shall be helped by several sources that opened my eyes: Hempel (1965, 1966); Lambert and Brittan (1992); Psillos (2002, 2007); Salmon (1990); Skow (2015); Weber, Van Bouwel, and De Vreese (2013); and Woodward (2011). Here, it is noteworthy that in most cases the explanatory models were presented to readers in the form of books (bulky volumes in several instances). The present chapter cannot accommodate an extensive and detailed discussion of these models with all the repercussions arising from them, and with all the interesting debates that they generated. Instead I attempt to present briefly the heart of the matter, these models' foremost and central notions, as well as two or three important critiques. Likewise, I shall endeavor to reduce to the minimum technical descriptions (e.g., a formal-logic presentation of logical arguments and proofs) so as to facilitate smooth reading of the surveys.

THE DEDUCTIVE-NOMOLOGICAL (D-N) MODEL

The main purpose of this model is to propose a format for a general scientific explanation for particular events. For example, Galileo's law of free fall can help explain why a stone that falls from a building of height *a* will hit the ground in *b* seconds. This model also suggests a way to explain empirical generalizations, for example, explaining the laws of Galileo and Kepler by Newtonian theory.

It seems to me that the best way to present the D-N model is by an illustration (the word *nomological* derives from the Greek *nomos*, meaning "law"). We shall apply once more the example of the free fall of bodies. Assume that David heard that in the first second an iron ball undergoes a fall of about five meters, and he wondered why this was so. The explanation, the answer to the question, is given by the D-N model as a logical argument: from certain (true) assumptions one infers through rules of logical deduction a (true) conclusion that describes the given phenomenon. Here is an example of a logical argument well-known to everyone:

Premises:

1. Every person is mortal
2. Socrates is a person

Conclusion: Socrates is mortal.

The D-N model has the following form of a logical argument:

Premises:

1. General laws, for example, Galileo's law: $S = 1/2gt^2$ (S is distance of fall; g is acceleration caused by gravitation; t is time of fall.)
2. Antecedent conditions, for example, $t = 1$ sec.

Conclusion: A description of the investigated phenomenon, for example, when $t = 1$ sec., $S = 1/2g = 4.91$ meters (which is about 5 m). (Note that the broken horizontal line below the premises signifies deduction.)

Thus, according to the answer given to David's question with the aid of the D-N model, it arises that the given phenomenon that he heard of is wholly explained, because a true conclusion is deduced from true premises of the explanation and because the conclusion well suits the description of the given phenomenon (i.e., the observation). The premise is commonly known as the *explanans* (the "explainer"), and the phenomenon under study is known as the *explanandum* (the "to be explained"). (Note here that Galileo's law is regarded as deriving from Newton's laws, perceived as fundamental and basic.)

As stated, the present explanatory model is constructed as a logical argument. This structure is based on the law in the *explanans* being expressed as a conditional sentence: "If . . . then. . . ." To clarify this, I shall again use an illustration. Let us examine the startle response, in which the dog Balak has responded to the sudden slamming of the door, and try to explain it by the D-N model. To offer an explanation of this kind, we shall first assume hypothetically that psychologists have discovered a law called "the law of the natural defense response": if *a sudden, strong stimulus* strikes the senses of any *land mammal* (*which lives on land*), then this creature will respond with a *natural defense response*. (The words in italic describe variables, categories, under which may be placed various values or examples.) And now we shall use this law to explain the above fact that Balak performed a startle response when the door slammed shut with a resounding *bang* caused by a strong gust of wind.

Premises:

1. The law of natural defense response: *If a strong and sudden stimulus strikes the senses of any land mammal (which lives on land), this creature will respond with a natural defense response.*
2. Antecedent conditions:
   a. The dog Balak is one example of *a land mammal* (*that lives on land*)
   b. The event "The door slammed shut with a resounding *bang* caused by a strong gust of wind" is a particular case of "a strong and sudden stimulus"

Conclusion: The dog Balak responded with a startle response, which is one response belonging to the category of *natural defense response*.

The explanation of the D-N model is based on a general law contained in the *premises*. In the first part of this law, in the *if* clause, lies a description of the situation in which the phenomenon in question occurred. In the second part of the law, the *then* clause, the part that describes the conclusion, lies a description of the observation that constitutes a particular example to be explained.

The present model is based on four important requirements, which determine logical and empirical matches (see Hempel, 1965, pp. 247–248):

1. The explanandum must be a logical consequence of the explanans; . . . for otherwise, the explanans would not constitute adequate grounds for the explanandum.
2. The explanans must contain general laws, and these must actually be required for derivation of the explanandum.
3. The explanans must have empirical content; that is, it must be capable at least in principle, of test by experiment or observation.
4. The sentences constituting the explanans must be true.

Two interesting implications arise from these requirements:

First, it is suggested that a distinction be made between a potential explanation, which meets the first three requirements, and the true explanation, which in addition to meeting the first three requirements also meets the fourth requirement, namely that the *explanans* be true.

Secondly, the fulfillment of these requirements also upholds the two demands for the D-N model. The first is that of *explanatory relevance*: the demand is that the occurrence of the phenomenon in question be expected as a result of its deduction logically from the given conditions. In this regard Hempel writes that satisfaction of this request leads to understanding why the phenomenon occurred: "The occurrence of the phenomenon *was to be expected*; and it is in this sense that the explanation enables us to *understand why* the phenomenon occurred" (1965, p. 317; italics in the original). The second demand is *testability*: the explanation has to meet an empirical test—namely the scientific explanation is able to pass empirical tests.

This model gave rise to much criticism. Some of it centered on problems stemming from the formulation of the model in the language of logic. I shall not discuss this aspect (see, e.g., Salmon, 1990). Another notable set of critiques emanated from this model's affecting to possess "the general characteristics of scientific explanation" (Hempel & Oppenheim, 1948, p. 136; it is worth noting here that the criticism appeared only about a decade after the publication of the paper in 1948):

1. The attempt to define a universal natural law is extremely problematic, and it is very hard to draw a line cutting across from a universal law to an accidental empirical generalization. (This situation is typical both of deterministic laws that appear in the present model and of probabilistic laws that appear in statistical models, which are described later.) This is a most thorny issue, and neither Hempel nor other researchers were able to propose a decisive criterion whereby it can be judged what a universal natural law is and what is nothing but an accidental generalization (e.g., Salmon, 1990). This

distinction is of supreme importance, because according to Hempel an explanation cannot be proposed based on an accidental generalization, but it can be proposed based on a universal natural law. We may look at the following example: assume that David has a moustache and a goatee, and the question arises as to why he has a moustache and a goatee. Someone suggests the following explanation: David belongs to the "Modern Literature Discussion group," and all members of this club have moustaches and goatees; therefore David has a moustache and a goatee. In light of this example, it is clear that Hempel's claim is correct. For comparison, let us look at the answer to the following simple question: Why is it easy to push a bookcase empty of books and impossible to move it from its place after it has been filled with philosophy books? We certainly will not accept an answer of this sort: this bookcase is extremely heavy because it is laden with the philosophical wisdom contained in the books of the greatest of philosophers from the age of Greece to our day; but we will accept an explanation resting on concepts of friction, weight, and gravity, concepts woven into relevant physical laws.

2. Despite the claim that the model is general, there was a need to develop models of explanation for other research areas: first, for phenomena characterized not as a particular event, which the D-N model addresses, but for events (phenomena) characterized as statistical. This subject was managed later with the development of appropriate statistical models in the framework of Hempel's scientific conception.
3. It emerged that the explanatory structure of the argument did not suit functional or teleological (purposeful) explanations or explanations based on social norms (see discussion in Rakover, 1990; Salmon, 1990). For example, how may one explain the fact that David stood on a chair to take a book off the top shelf in his library? It seems that no law or universal theory exists from which this behavior of David can be inferred. In this connection it became evident that it is very hard to offer explanations in psychology and in history with the models developed by Hempel (see Dray, 1966). That is, not only is the application of the D-N model to these domains complicated, but also application of the statistical models seems awkward and a misfit, and other kinds of explanation are needed that are not on the lines of a deductive or inductive argument (e.g., Mahajan, 2011). Similarly, it would be very hard to apply the present model to explain the following phenomenon: David stopped his car the moment the traffic light before him changed from green to red. There is no law or sound empirical generalization that can accommodate this event, simply because David took the path of following the traffic rule that the state obliges drivers to obey (otherwise they will be penalized). This, then, is an arbitrary rule intended to organize most efficiently the flow of traffic with a minimum of road accidents.

4. The D-N model is hard-pressed to deal with causal phenomena, namely events that intuitively require an explanation by appeal to causality. Here I shall discuss briefly several examples considered classic, and that show that explanations suited to the D-N model can be constructed, but clearly these explanations are flawed. It emerges from them that neither the D-N model nor—as we shall soon see—Hempel's proposed statistical models constitute a general explanation pattern that characterizes scientific explanations in all fields.

The Flagpole and the Shadow

With this illustration, attributed to Bromberger (see Salmon, 1990), it seems very easy to answer the following question by means of the D-N model: given a flagpole of length *a* standing in the parade ground, and a sunray at a certain angle (say 45°), why is the shadow's length *b*? The answer, the explanation, will be given through geometrical laws of light, of trigonometry (the appropriate angle and the flagpole's length), from which it may be deduced that the length of the shadow cast by the flagpole onto the parade ground is *b*. So far, the explanation for the shadow's length has followed faithfully the present explanatory model, and there is no problem here. However, under the same conditions the question can be reversed: Why is the flagpole's height *a*? Again, there is no problem of calculation in offering an answer. But clearly, the explanation in this case does not hold! Why not? Once more, because while it is clear that the flagpole is the cause of the shadow, it is extremely difficult to imagine that the shadow is the cause of the height of the flagpole. And even though the explanation seems to us incorrect and unacceptable, the answer obtained fits precisely the pattern of the D-N model. An illustration like this, with the same conclusion as regards the D-N model, concerns similar questions about the movement of a pendulum expressed as the "pendulum law," $t = 2\Pi\sqrt{l}/g$. When *t*, the time of the period of the pendulum, is explained by means of *l*, the pendulum's length, the explanation seems causally logical; but when the pendulum's length is calculated according the period, the explanation seems unreasonable causally.

Addressing the causality problem, Hempel (1965) proposed a distinction between two kinds of law (see also the discussion in Psillos, 2002): (1) laws of coexistence. These are laws that connect different variables in the framework of a single function, for example, the pendulum law, which may be given a general expression of the following kind: $Y = f(K, X)$, where *K* signifies a certain constant and *X* and *Y* stand for variables (in the case of the pendulum law, *t* was set in *Y* and *l* in *X*). These laws do not describe changes in time (while the usual conception is that the cause precedes the result in time). (2) Laws of succession. These laws describe development in time, for example, Galileo's law above. In Hempel's view, although causal explanations have a structure of the D-N model, not all explanations are causal. Causal explanations are based on laws of succession. This approach gave rise to controversy. For example, Psillos suggested that even regarding the Boyle–Marriote law of an ideal gas, considered a law of coexistence and formulating the equation $PV = K$, where P is pressure, V is volume, and K is constant, we will ascribe causally the reduction of the gas volume to a rise in pressure as a result of the compression of the gas. (In this case, the rise in pressure will cause a rise in the gas temperature, according the Charles–Gay-Lussac law that the gas volume equals temperature × constant.)

The Barometer

Noticing that the barometer has fallen sharply, we will predict that a storm is about to batter the region. Because this observation recurs many times and at different places on earth, a barometer law may be proposed: whenever barometric pressure falls sharply, a storm is due. Based on this, many cases where the barometer falls steeply and a storm takes place may be explained by the D-N model. Yet it is inconceivable that the fall of pressure in the barometer caused the storm. We know that the two events—fall in barometric pressure and occurrence of the storm—are created as a result of atmospheric changes (fall in atmospheric pressure, which gives rise to strong gusts of air).

In actual fact, this example points up two weaknesses of the D-N model. On the one hand, it struggles with causal explanations; on the other hand, the example highlights an important difference between prediction and explanation. It is possible to suggest according to the present model a good and suitable prediction, without this prediction being accompanied by a suitable explanation. This example, therefore, runs counter to the explanation–prediction symmetry thesis that underpins the D-N model. According to this thesis, the conclusion arising from the premises is a prediction, and it also constitutes a basis for the explanation of the D-N model (the phenomenon is expected, given the premises of the present explanatory model). Hence, according to the present model, an excellent prediction may be proposed without its being accompanied by a good explanation. Furthermore, researchers have drawn attention to reverse cases also, that is, to instances that offer reasonable and conceivable explanations without these being capable of offering good predictions. For example, the theory of evolution can explain nicely the development of different species over lengthy periods in the ancient past, but will find it very hard to predict future development. The same may be said of many theories in the psychology of personality. For example, psychoanalysis can explain certain behavioral disturbances by an appeal to the activity of different complexes in childhood, but to the best of my knowledge this theory will find it hard to predict future development on the basis of these same observed complexes.

Lastly, I set forth an illustration (after Salmon, 1990) showing that it is possible to offer a prediction, an explanation, that does not have recourse to a law or theory as required by the present model. For example, I go to the coffee room in the psychology department, make coffee, open the refrigerator, and pour milk into the coffee. When I sip the coffee, it tastes abominable. The milk has soured. I check, and sure enough, the milk has gone off, and it tastes awful. I predict that anyone who uses the milk will get a taste of the sickening sourness, so I put a sticker on the milk container declaring, "The milk is sour. Don't use it."

STATISTICAL MODELS FOR GIVING EXPLANATIONS

In this section I concentrate on three important statistical models: two proposed by Hempel, namely the Deductive-Statistical (D-S) model and the Inductive-Statistical (I-S) model. (Hempel's three models—these two and the D-N model—are known as Covering-Law models, because according to them the phenomenon to be explained is contained in or covered by the law [theory] that appears in the *explanans*.) And thirdly, a model proposed by Salmon: the Statistical-Relevance (S-R) model (e.g., Salmon, 1990; Salmon et al., 1971). I shall begin with a short survey of Hempel's two models and then describe briefly Salmon's model, which was intended to replace the I-S model.

The Deductive-Statistical (D-S) Model and the Inductive-Statistical (I-S) Model

These two models are founded on the basic ideas of the D-N model: they have an argumentative structure (a deductive argument in the first, an inductive argument in the second), and they have justification as providers of explanations because the occurrence of the *explanandum*, imparted by the *explanans*, can be expected with full or very high probability.

The D-S Model

This model is closely similar to the D-N model. The basic differences between the two are these: while in the D-N model there appears in the *explanans* a law, theory, concerning the occurrence of events, in the D-S model there appears a probabilistic law, theory, concerning the probability of the occurrence of events; while the D-N model treats the explanation of isolated events and also empirical generalizations, the D-S model treats only the explanation of probabilistic empirical generalizations—namely, it is meant to explain a probabilistic generalization by means of an extensive probabilistic theory or law. As in the D-N model, the statistical generalization is explained by its being inferred deductively (mathematically) from the *explanans*, that is, from the model's premises, which include at least one probabilistic law, a probabilistic theory, of the most general kind. As an illustration taken from psychology, one may suggest the explanation of the empirical generalization of the learning curve, according to which the probability of a response increases as a function of the number of trials, by the theory of Estes (1950) called the Stimulus Sampling Theory, and based on an associative connection between the stimulus and the response.

The I-S Model

The essential differences between this model and the D-S model are these: the I-S model is meant to explain the occurrence of a particular event by showing that this occurrence is expected at an extremely high probability (close to 1); inference of the probability of occurrence of the event from the *explanans* is not deductive but inductive. So it may be said that the *explanans* irradiates onto the *explanandum* the extremely high probability that the given event will indeed occur. The requirement of high probability is of great importance: if, for example, we say that the probability that event *X* will occur is low (e.g., 0.15), according to the present model we may say that with extremely high probability (0.85) event *X* will *not* occur. Here is a fairly simple example:

Premises:

1. A statistical law: The law of smoking—for every person, if he/she smokes thirty cigarettes or more a day, then his/her chances of contracting lung cancer are very high.
2. Antecedent condition: David smokes two packs (forty cigarettes) a day.

==

Conclusion: The chances that David will contract lung cancer are very high.

The explanation for David's illness is given by an appeal to the "smoking law": according to this law, it was expected with extremely high probability that David would fall ill with cancer. (Note that Hempel takes the double line separating the premises from the conclusions as an inductive, and not a deductive, inference, as in the D-N model. See above.)

The I-S model raised several problems (in part handled by Hempel, 1965, himself). One of them, which does not touch on explanation based on a deductive argument, but is possible in any explanation based on inductive argumentation, is this: addition of information to the premises (to the *explanans*) is likely to change the inductive conclusion. To substantiate this, we may add to the cigarette-smoking illustration new information expressed in the following law: the law of cancer-immunization. Suppose that David has been given an injection to immunize him against lung cancer (an immunization that perhaps will be developed in the future) so that the probability of his falling ill with lung cancer is negligible—that is, the probability that he will be healthy is very high.

Considering the added relevant information, a problem is indeed created, namely, the existence of two I-S explanations, with extremely high probability, giving contrasting predictions about the same person: he will and he will not fall ill. According to the smoking law, we will predict with very high probability that David will contract cancer. But according to the cancer-immunization law, we will predict that David will not contract cancer, also with very high probability. This is a troublesome, ambiguous situation, which Hempel called the *ambiguity of I-S explanation.* To eliminate it, Hempel proposed a modification to the model: the *explanans* must satisfy the *requirement of maximal specificity* (RMS). Namely, the *explanans* must include a law or theory based on all known relevant information. Now it emerges that the smoking law does not satisfy the RMS requirement (because the explanation did not take into account all the information known on the subject), while the cancer-immunization law does satisfy it; therefore, we must accept the statistical prediction that with extremely high probability David will remain healthy, and reject the contrary prediction.

The I-S model gave rise to some criticism, which eventually generated the development of alternative explanatory models (see below). Here I shall touch briefly on four criticisms. The first holds that the requirement of a statistical law presents problems very similar to the requirement of a deterministic law in the D-N model and is connected to the question of what distinguishes a law from an accidental generalization; the second states that RMS creates tricky problems of division into groups, reference classes, to which the issue of the given law applies (e.g., a group of heavy smokers as against light smokers and everyone else) (see discussion in Humphreys, 1968; Salmon, 1990).

The third criticism presents a counterexample, which attests that with the I-S model the *explanans* is liable to contain information that is not relevant to the *explanandum.*

1. People who have a cold and take vitamin C have a 0.80 probability of getting better within a week;
2. David has a cold and has taken vitamin C.

Conclusion, prediction: David is likely to recover within a week with a probability of 0.80.

The interesting point in this example, as Salmon (1990, p. 50) points out, is that although this explanation seems legitimate—namely, it is an example that fits the format of the present model—the model cannot filter out irrelevant information: people recover within a week from a cold whether they take vitamin C or not. That is, vitamin C is not relevant to curing a cold.

The fourth criticism also presents a counterexample, this one showing that there are legitimate statistical explanations even though they do not match the I-S model (the model is not a sufficient condition for a statistical explanation).

1. People who have not had treatment for latent syphilis have a 0.25 probability of developing paresis (brain paralysis).
2. David has not had treatment for latent syphilis.

==

Conclusion, prediction: David is likely to develop paresis with a probability of 0.25.

The interesting point in this example is that although the explanation is legitimate, it does not fit the I-S model because the probability irradiated from the *explanans* onto the *explanandum* is very low.

As may be seen, Hempel's models (D-N model and I-S model) generated severe criticism, which centered on the following questions: Are these models general, applicable to all areas of science? Is a scientific explanation based on an argument (deductive, inductive)? Is a statistical explanation analogous to the D-N explanation? Does a statistical explanation necessitate the use of high probability (close to 1)? As a result, several researchers proposed alternative models, one of which I shall discuss briefly here.

The Statistical-Relevance (S-R) Model

The ideas underlying this model are as follows: for a statistical explanation, there is no structure of a deductive or inductive argument, as Hempel suggested; the probability irradiated from the *explanans* onto the *explanandum* need not be high, attesting to the degree of its being valuable (see Salmon, 1984, 1990; Salmon et al., 1971). I shall use the following well-known illustration to clarify this model.

The probability (represented as P_r) that a person will recover (R) from an infection by streptococcus (S) after given penicillin (P_n) medication is expressed thus: $P_r(R/S\&P_n)$. Assuming that the penicillin did indeed help to cure the infection, we may say that $P_r(R/S\&P_n) > P_r(R/S)$—that is, the probability of recovery under the condition of having the penicillin (the so-called posterior probability) is *greater* than the probability of recovery without having the penicillin (the so-called prior probability). In this case we say that giving penicillin has statistical relevance for recovery from the streptococcus infection.

In general, Salmon suggests that for reference class (F) factor C explains event E when

1. $P_r(E/F\&C) > P_r(E/F)$ (in this case C is positively relevant to E)
2. $P_r(E/F\&C) < P_r(E/F)$ (in this case C is negatively relevant to E)

Generally, Salmon holds that factor C is relevant statistically to the occurrence of E when there is a difference between the posterior probability and the prior probability ($P_r[E/F\&C] \neq P_r[E/F]$). The prior probability of event E ($P_r[E/F]$) is determined by the frequency of E in its reference class (F). (For example, the streptococcus illustration refers to the probability of spontaneous recovery in a group, a class, of people infected by this microbe.) The posterior probability of event E ($P_r[E/F\&C]$) is determined by the frequency of E in its reference class *on condition* that factor C has been activated in this group. (For example, the probability of recovering from streptococcus after the patient has been given penicillin.)

As may be seen from the account so far, in contrast to the I-S model, Salmon's explanatory model is not based on a structure of argument (deductive or inductive); the explanation is based on two kinds of probability instead of one (extremely high), which appears in the I-S model; in the S-R model the size of the probabilities is of no importance (they need not be large, close to 1)—what is important is that the inequality $P_r(E/F\&C) \neq P_r(E/F)$ exists. For example, even if $P_r(E/F) = .05$, which is a very low probability (and according to the I-S model what we have in fact is a very high probability [.95] that E will not occur), factor C will be considered relevant when $P_r(E/F\&C) = .25$ (which is also a very low probability according the I-S model); however, a probability that supports factor C explains E, because by the present model the equation ($P_r[E/F\&C] = .25$) > ($P_r[E/F] = .05$) is upheld.

On the face of it, if the relation $P_r(E/F\&C) > P_r(E/F)$ exists (the relation that determines that the probability of occurrence of E under conditions F and C is greater than its occurrence under the condition of F alone), we may think that C constitutes a cause for the occurrence of event E. But this conclusion is too hasty, as may be seen from the following analysis. Let us scrutinize once more the illustration of the barometer and the storm. We shall denote the fall on the barometer as (B), the storm as (S), and the fall in atmospheric pressure as (A). That the fall on the barometer predicts very well the advent of the storm may be written in terms of the present model as follows: $P_r(S/B) > P_r(S)$. But is this inequality correct? And may one infer from this that the fall on the barometer is the cause of a storm? The answer to both questions is negative. Why? Because the fall in atmospheric pressure A is the cause common to B and S. That is, it will be correct to write the following probabilistic *equation*: $P_r(S/A\&B) = P_r(S/A)$, namely the barometric change is not relevant to the explanation for the occurrence of the storm.

So treatment of the concept of causality by means of the statistical tools offered by the model is not so simple. Let us take yet another look at the streptococcus case. The probability that a person will recover from the strep infection, given the penicillin medication, was, as stated, expressed thus: $P_r(R/S\&P_n)$. But it can happen that a person will be infected by a microbe that cancels the effect of the penicillin (T). In that case an *equation* will again come about: $P_r(R/S\&P_n\&T) = P_r(R/S\&T)$, meaning in this case (where the individual has been infected by an anti-penicillin microbe, T) the penicillin medication becomes irrelevant, and factor T is perceived as the buffering factor that screens off R from P_n.

In short, it arises that for us to be able to infer from the relation $P_r(E/F\&C) > P_r(E/F)$ the conclusion that C constitutes a cause of E, we must ensure that additional conditions do not exist that will hinder the realization of this conclusion, such as the screen-off factor that we discussed in the streptococcus example.

It seems that the S-R model and the illustrations we have analyzed under its terms may well suit the methodology of conducting experiments in psychology. In this connection, Salmon (1990, p. 59) writes the following:

> What is crucial for statistical explanation, I claimed, is not how probable the *explanans* renders the *explanandum*, but rather, whether the facts cited in the *explanans make a difference to the probability of the explanandum*. . . . By comparing the outcomes in an experimental group (the members of which receive the treatment in question) with those of a control group (the members of which do not receive that treatment), we procure evidence concerning the effectiveness of the treatment.

To substantiate this matter, we shall denote treatment in the experimental group (T) (e.g., the treatment gives positive reinforcement in an experiment on learning the appropriate response (R) in a T-shaped maze, where the reinforcement is given in the target box on the right-hand side). If indeed the treatment positively affects the response, we shall obtain the following inequality: $P_r(R/T) > P_r(R)$. In this case we shall be able to propose an explanation whereby the treatment (the reinforcement) is relevant statistically: it has increased the probability of the response's (R) appearance in the experimental group above the probability of its appearance in the control group.

Furthermore, it turns out that the treatment of the present model with the screen-off factor also finds expression in a widely accepted statistical analysis, called Partial correlation (see, e.g., Guilford, 1965). For example, researchers found an extremely high correlation between A and B (r_{ab} = .8), and also a high correlation between A and C (r_{ac} = .78); however, these researchers posited that C was responsible for the high correlation between

A and *B*. To test this assumption, they calculated the correlation between *A* and *B* while holding *C* constant. If the researchers' assumption is correct, this correlation (i.e., when *C* is held constant) will have a much smaller value than the previous correlation, and indeed this was obtained ($r_{ab.c} = .03$ vs. $r_{ab} = .8$). In terms of the present explanatory model, when *C* was held constant, a partition was placed between *A* and *B*. Still, several important differences between this model and accepted research methodology are noteworthy.

Compared with what Salmon's model requires, the methodology of experiment emphasizes several more items of great importance. We shall look at two differences. First, research methodology also requires a statistical test of this question: Is the gap ($P_r[R/T] - P_r[R] > 0$) an accidental result or not? (For example, does the difference arise from accidental errors of sampling the participants and from measurement mistakes?) Secondly, the methodology requires that the experimental design be justified by a theory from which arises the prediction that giving *T* will indeed improve the response (the probability of its appearance will increase).

CAUSAL MODELS FOR GIVING EXPLANATIONS

As may be seen from the brief accounts of Hempel's and Salmon's models, the latter could hardly handle explanations connected to the idea of causality, that is, to the intuitive, almost natural, explanation that event *E* occurred because another event *C* appeared before event *E* and caused its occurrence (e.g., in a lab experiment the independent variable was operated first, and only afterward did the participant respond). In this section I briefly survey three models of explanation based on the idea of causality. The first is the Causal-Mechanical (C-M) model of Salmon (1984, 1990). The second is the Manipulationist (Interventionist) model of Woodward (2003); the third is the Kairetic model of Strevens (2008b).

The Causal-Mechanical (C-M) Model

Salmon's motivation to set aside his earlier model, the S-R model, and begin to develop the present model of explanation is set forth frankly in the preface to his book (Salmon, 1984, x–xi):

> This book had its real beginnings when, around 1973, I began to recognize the serious shortcomings and limitations of the statistical-relevance model of scientific explanation—as propounded in (Salmon et. al., 1971)—and the need to come to grips with the explanatory force of scientific theories. . . . As chapters 5–8 of the present book will show, I now believe that causal explanation is the

> key to theoretical explanation. No adequate account of theoretical explanation could be developed without extensive work on causality itself.

Salmon assumes that the world is set on a causal mechanism, a causal structure, and the role of explanation is to identify this structure, mechanism. The present explanatory model rests on two basic concepts of utmost importance: causal process and causal interaction. To best clarify the model and its terms, I resort to the following illustration (after Salmon, 1978). Assume that on August 6, 1945, Hashimoto was two and a half kilometers away from the atomic bomb explosion that hit Hiroshima, and after a short time he contracted leukemia. How can we explain this unfortunate event? Given the C-M model we will suggest that a causal process occurred; namely, powerful radioactivity emanating from the nuclear bomb ("Little Boy" was the name of the bomb dropped from the US B-29 bomber called *Enola Gay*) spread radially on all sides, and soon traversed the distance between the explosion and the place where Hashimoto was. The radiation attacked Hashimoto's blood cells and caused a grave disturbance to their normal functioning in the cell proliferation mechanism, so that it began to multiply unstopped and to create malignant growths—that is, leukemia was created in Hashimoto's blood. And now we may examine how this short explanation accords with the explanatory scheme proposed by Salmon.

As stated, in Salmon's opinion, a causal explanation must provide us with understanding by giving an answer to the *why* question. The answer is provided by a description of causal processes behind the event under study. In the C-M model the causal process is an important concept: it is a continuous physical process in space and time, for example, the flight of a bird in the sky, the movement of a car on the highway, a pool ball rolling across the pool table surface. In the Hashimoto example the causal process is radioactivity (i.e., gamma waves and x-rays), which spread radially and is transmitted between the atomic blast and Hashimoto. This is a continuous, not a broken, process in space/time. The causal process creates a causal interaction that is described as an event created from the crossing, or intersecting, of two causal processes, where the change in both continues after the interaction has ended. An example is the collision of two pool balls: the first, which moves, speeds toward the second, which is stationary; after the interaction the first ball changes its direction and speed, and the second begins to move in a direction determined by the strike angle of the first ball against it. In the Hashimoto example, the damage caused by the radioactivity to the blood cells gave rise to continuous destruction of the cell and its components (damage to the DNA spiral), taking it even to the growth of cancerous cells. What typifies a causal process is its ability to transmit a causal effect, sign, mark, which remains even after the interaction of one process with another has ended (i.e., after the interaction the two processes preserve the change over time). For example, when two cars collide, the twisting of the fenders on both is a kind of mark, a sign transmitted by the causal process that endures even after the causal interaction (the collision) that created this mark has ended. The interaction creates the mark, which is the product of the change in the structure of the process. Note that a process marked by the interaction with a second process is capable of passing the mark on. On this Salmon writes: "A single intervention at one point in the process transforms it in a way that persists from that point on" (1984, p. 142). The foregoing can be summed up thus: if causal process *P* characterized by quality *Q* undergoes an interaction with another process, such that quality *Q* changes into *Q'*, then henceforward process *P* will transmit in the next interaction quality *Q'* (naturally, the quality [*R*] of the second process also undergoes change as a result of the interaction: from *R* to *R'*).

Some criticism was voiced against Salmon's C-M model, some of which Salmon tried to rebuff (see Psillos, 2002, 2007; Salmon, 1997; Woodward, 2011). Here, I shall briefly describe two critiques, one claiming that the C-M model cannot distinguish causal factors relevant to the explanation of the given phenomenon from those not relevant; the other claiming that the present model encounters cardinal problems the moment one tries to apply it to complex systems.

Relevance

David chalked the tip of his cue, and then used it to hit the cue ball, which rolled fast over the table and struck the eight ball. The result was that the eight ball was deposited in the corner pocket and the cue ball changed the direction of its movement, hit the cushion, bounced off it, and came to a halt some distance from it. Furthermore, the cue striking the cue ball left a spot of chalk on it, which then marked the eight ball as well. In terms of the C-M model, the strike of the cue against the cue ball (i.e., the interaction between the cue and the cue ball) was the product of a causal process that transmitted to the cue ball the mark (the chalk spot), which passed further on to the eight ball as well! But is this causal process of transmission of the chalk mark the proper explanation, the relevant explanation, for the movement of the cue ball to the eight ball? And the answer is clear: no it is not. To explain the movement of the balls, calculations must be made that use the laws of movement detailed in Newtonian theory (having to do with mass, velocity, direction, laws of preservation, and the like)—this is the relevant theory to explain the balls' movement over the pool table. What arises from this example is an argument stating that it is hard for the C-M model to distinguish a relevant causal process for an explanation of the given phenomenon (the calculations underlying Newtonian theory) from a nonrelevant process (transmission of the chalk mark).

Complex Systems

It is difficult to see how the C-M model may be applied to complex systems based on proliferation of causal processes and interactions, doubled and tripled many times over, such as biological and psychological systems. There it would be prodigiously complicated to trace each and every causal process. For example, how may we explain the fact that David stopped his car at a red light? Let us suppose that we can follow the different interactions of the red light from the retina of the eye through all the neurophysiological cycles in the brain, including innervations of the motor system, culminating in the pressure of the foot on the brake pedal. We still cannot supply an answer regarding to the conscious significance associated with the braking action, the purpose of the foot's pressure on the brakes (the car can be stopped by other means too); we still cannot explain why a scene in a movie where a red traffic light goes on at an intersection does not spark in David even the slightest hint of the causal processes described above. Even in the simplest physical systems, the story is not at all simple. For example, how may one trace the causal process of every single molecule of gas in a container as the pressure on the gas changes?

The Manipulationist-Interventionist (M-I) Model

Woodward's (2003) M-I model is also based on the idea common to the other causal models reviewed here, namely, explanations founded on causal factors. The basic difference among these three models lies in the kind and nature of the causal theory underlying the explanatory model. In essence, the M-I model may be called a "counterfactual model of explanation." The basic question that this model strives to answer is this: What would have happened if things had been different? (Let's call it the *what-if* question for short.) How might a possible causal change (probably because of a certain manipulation/intervention) have affected the occurrence of the phenomenon under study? Look at this very simple illustration. David flicked the light switch, and the room was flooded with light. Now ask, if someone had intervened and stopped the movement of David's arm, what would have happened? Leaving aside David's response, the answer is that the room would have remained in darkness. Or in another illustration, what would have happened if David had not taken the penicillin medication against pneumonia? While the causal explanation according to Salmon (1984) is based on a description of a physical process in space/time, including the interactions of this process,

Woodward's explanation is based on two important elements: manipulation/intervention and counterfactual conditions.

In Woodward's view, a model of explanation must answer the *what-if* question. He writes (2003, p. 11):

> an explanation ought to be such that it can be used to answer what I call a *what-if-things-had-been-different question*: the explanation must enable us to see what sort of difference it would have made for the explanandum if the factors cited in the explanans had been different in various possible ways.

Causal explanations are connected to counterfactual situations, for which manipulation/intervention is impossible in practice. For example, Woodward proposes that if it had been possible to prevent the collision of the great asteroid in the age of the dinosaurs, their annihilation (and the annihilation of other species) would have been prevented. This counterfactual circumstance well supports the hypothesis that what annihilated dinosaurs in one fell swoop was the collision of a huge asteroid with planet earth, which caused climatic changes so severe that they resulted in the extermination of these animals.

According to the M-I model, X is the cause of Y if the manipulation of X causes a change in Y. For example, if David pressed the electricity switch to *on*, the room would be flooded with electric light. Generally, a certain intervention I creates a causal process: $I \rightarrow X \rightarrow Y$, where the arrows signify a causal effect. (Cause $I \rightarrow X$ is not identical with Cause $X \rightarrow Y$.) By the present model, X explains Y causally if a change in X causes a change in Y. But here two more requirements enter: (a) the change in X will be caused only by I and not by some other variable Z that is effected by I and affects Y; (b) the change in I does not directly affect Y.

According to the M-I model, manipulations/interventions are closely connected to invariant generalizations, which facilitate them; the empirical generalization that connects variable X to variable Y ($Y = f[X]$) facilitates making a change in the causal variable (X), which will cause a change in the given phenomenon (Y). Woodward (2003, p. 15) writes: "The notion of invariance is closely related to the notion of intervention."

Invariant empirical generalizations differ from laws of nature principally because they are exceptional and not universal like the laws of nature, which apply everywhere and are true at all times. These generalizations serve as extremely important aids in giving explanations in the social sciences such as psychology. As a possible example of an invariant empirical generalization in psychology, consider the following: "An animal fearing a threatening source will tend to increase the distance between itself and that source." This generalization will explain many behaviors—for example, the postman will run away from a barking dog that bares its teeth. But this generalization has exceptions: in several cases the frightened creature will not run off but will freeze in place (e.g., when the fear is great and sudden); and in several other cases the frightened creature will attack the threatening source (e.g., a lioness will go out to protect her cubs from threatening predators).

The connection between manipulation/intervention and invariant generalizations is nicely illustrated in the case of the well-known generalization that a fall on the barometer predicts a storm. It can be shown that this connection is not causal if we consider a counterfactual situation where if we perform a manipulation on the barometer itself so that it indicates calm weather, the storm will not be stopped. (The point here is that the cause of both the storm and the fall on the barometer is the steep drop in atmospheric pressure.)

What is the advantage of this conception of Woodward's? First, the manipulation need not be wholly dependent on a person's deeds, because causal intervention is likely to occur naturally in the world. Moreover, Woodward expressed the causal relation between *X* and *Y* as a relation between variables, where *X* is likely to obtain different values, therefore also *Y*. For example, in the case of switching on the light in the room, *X* obtains two values—on or off—and *Y* obtains two values—there is or there is not electric illumination in the room. Changing the switch from *off* to *on* will cause a change from darkness to light.

Secondly, by this approach we do not encounter problems that confront other approaches to causality. For example, according to the approach proposed by Hume in his day, called the regularity theory of causation, everything we see in nature is a recurrent succession of events: event *Y* appears after event *X*, for example, day/sunlight that appears after night/darkness, or whenever I open the palm of my hand, the pen (or anything else I'm holding) falls to the floor, and thousands more examples. Hume's approach may come across several instances that contradict logic. For example, let us look at the regularity *X* causes *Y*, and set in *X* as a particular case this event: David regularly takes prophylactic pills against pregnancy; and set in *Y* as a particular case this event: David does not get pregnant. So indeed we can see that in a thousand cases where David swallows a pill against pregnancy, he does not become pregnant. This example is a favorite of philosophers (see Salmon, 1990) because, among other things, it's funny. Will Woodward's approach similarly come up against this entertaining obstacle? It will not. If we apply the *what-if* question in the present case, namely, if we ask what would happen if David did not take the prophylactic pill, a crystal clear answer will be obtained. (In the present case *X* has two values—takes, does not take a pill against pregnancy—and *Y* also has two values—gets, does not get pregnant. The absence of an effect of *X* on *Y* is revealed when the values of *X* are changed and the same value is obtained in *Y*: does not get pregnant.)

The M-I model, like the other explanatory models, gave rise to some criticism. I shall concentrate very briefly on the following two instances. First, it can be seen (and Woodward himself underlines this in his book) that the model is firmly linked to the conduct of experiments—in fact, the *what-if* question is an invitation to design an experiment: Considering the invariant empirical generalization (*Response* = *f*[*Stimulus*]), will a response arising from a certain change in stimulus be obtained? The problem with this outlook is that this approach does not harmonize with the experimental work in which researchers in psychology engage. Very few experiments in psychology are based empirically on invariant generalizations (or what is called in psychology "laws," for example, the "matching law," which proposes equality between the proportion of reinforcements that response R_1 gets out of two, and the proportion of R_1 out of the two responses). In most cases psychologists run experiments to refute or support one hypothesis against another, for example, a series of experiments concerning the question whether memory is based on one memory storage or two, or whether the highly efficient face perception is grounded in a particular mechanism specific to the perception of faces or originates in wide-ranging daily learning, lasting years upon years from the day of the person's birth.

Secondly, Saatsi and Pexton (2013) put forward the argument that Woodward's (2003) M-I model yields a mismatch between the counterfactual element and the causal element. It emerges that the explanation arising from the viewpoint of the counterfactual situation covers noncausal explanations as well. They point out, for example, that geometric explanations, which are deemed noncausal, may be analyzed according to counterfactual situations (the *what-if* question). Note here that several researchers maintain that all explanations given to particular events are causal (e.g., Skow, 2014). I personally do not believe that all explanations for specific events are causal. Here are two illustrations that support this statement.

1. In chapter 1, I presented an illustration on the conditions under which a square surface will pass through a round hole. The answer, as stated, is when the diagonal of the square is smaller than the diameter of the circle. This is an answer from the world of plane geometry, and it seems disconnected from a causal explanation.
2. Suppose one asks why out of a thousand people at the park concert in Tel Aviv, only 7 percent (or so) have IQ above (>) 135. The answer is based on the known fact that the distribution of IQ is Gaussian, in which 7 percent of it has IQ > 135, and the assumption that the thousand people at the park represent a random sample of this distribution. This is an acceptable statistical explanation, which is hard to connect to causality. Of course, one can suggest several factors such as genetics, education, nutrition, and so on, which may causally explain how their unified complex effect brings about the IQ distribution. But even if one grants this explanation for the

generation of the IQ distribution, the explanation for the 7 percent of the thousand people who have IQ > 135 will still remain statistical.

The Kairetic Model

One may say that the explanatory model proposed by Strevens (2008b) in his large volume is essentially an attempt to answer the basic question of what a relevant causal explanation is. Salmon (1984) set the causal explanation on a causal process, which transmits a certain mark across space and time; Woodward (2003) set the causal explanation on counterfactual situations. Strevens, however, strove to develop a methodology, a method, whereby it would be possible to build a model based on difference-making causal factors to explain the occurrence of the phenomenon under study. In other words, finding a difference-making causal factor for why/how the given phenomenon occurred would answer the causal relevance question. I believe that the best way to clarify Strevens's approach is by an illustration. I select a historical one, which Strevens presents through the Kairetic model (see Strevens, 2004, 2008b). (The word *kairetic* derives from the ancient Greek *kairos*, meaning "crucial moment," and it serves to indicate a critical difference-making factor. See footnote 1 in Strevens [2008b].) The illustration concerns a historical event that occurred toward the end of Tsarist Russia, namely the murder of the monk Rasputin, who enjoyed magical powers.

Grigori Yefimovitch Rasputin Novikh, born 1869, was renowned as a monk possessed of supernatural powers (healing and clairvoyance). He was close to the imperial court of Tsar Nikolai II, and there acquired a firm position, to the extent of exerting powerful influence on political appointments. Rasputin made not a few enemies, and late in 1916 he was invited by a ring of conspirators/murderers (among them Prince Felix Yussupov and the Grand Duke Dimitri Pavlovich) to a banquet at Yussupov's stately home. First he was offered cakes and wine poisoned with cyanide, but the toxin did not kill him. So the plotters shot him, but Rasputin managed to escape into the residence's grounds, where they caught him, bound him hand and foot, and dropped him through a hole in the ice into the River Neva. An autopsy showed that Rasputin had drowned to death. (The man was apparently incredibly strong: he had recovered from an earlier attempt to kill him by stabbing him in the stomach.)

Strevens uses this example to answer the question of relevance, namely, finding the difference-making cause. Three important factors are involved in the monk Rasputin's death: (a) poisoning, (b) shooting, (c) drowning. (Here Strevens notes in addition to these three the action of a large number of other factors that might have been involved in Rasputin's death, such as the length of his beard and the gravitational pull of Mars. But it is quite evident that these factors are not relevant to Rasputin's death, so we discard them routinely and automatically. See Strevens, 2004.) By means of the so-called eliminative procedure, Strevens offers a way to eliminate the factors that are not difference making for the occurrence of the given phenomenon. Poisoning, it arises, was not the factor that killed Rasputin, because he simply did not die as a result of the cyanide in the time span in which he was expected to expire. More or less the same may be said about the shooting: the bullets did not kill Rasputin, who fled into the garden. Therefore, the cause of the monk's death was drowning, which thus is the relevant, difference-making, causal factor that killed Rasputin. It was the causal element out of all the factors involved in the attempt on Rasputin's life that made a difference as to whether he died or not.

Now, through this illustration it will be easier for me to explain the Kairetic model. The basic premise is that to answer why/how a certain event *e* occurred (e.g., Rasputin's death), a network of causal factors is involved, for example, the three factors surrounding Rasputin's death and a large number of others, which are not even referred to in the account of the monk's demise. The question before the Kairetic model is how to choose the difference-making factors out of this array. To succeed, Strevens developed the eliminative procedure that is operated on an explanatory model, which for convenience I shall call the primary explanatory model (*M*), based on several statements/sentences, which causally entail a description of the phenomenon under study (*e*). In other words, Rasputin's death is causally entailed from a number of laws concerning cyanide poisoning, grievous injury by gunfire, and drowning, and a number of actual situations where Rasputin was poisoned, shot, and drowned in a river. The eliminative procedure results in a new model (*M'*), which proposes a causal explanation for the phenomenon at hand. This model, *M'*, consists solely of the laws and situations that cannot be disregarded, for their elimination would result in our being unable to explain causally the monk's death. The model obtained in the present case, *M'*, explains Rasputin's death by means of the law that concerns drowning and the factual situation that Rasputin drowned in a river. Strevens describes the Kairetic procedure:

> The difference-making parts of a causal model are those parts that play an essential role in the entailment, meaning roughly that, if they were to be removed from the model, it would no longer entail *e*'s occurrence, or it would not do so in the right sort of way. This is what I call the *eliminative procedure*: remove as many pieces as you can from a model M for an event *e* without invalidating the entailment of *e*. Everything that remains in the model is a difference-maker for *e*. (2008b, p. 70)

In reference to the Rasputin illustration, Strevens compares his approach with two other criteria intended to test for causal relevance, that is, for causal difference making to the event's occurrence, the phenomenon under study. The first is the probability criterion developed by Salmon et al. (1971). By this criterion it arises, for example, that the gravitational effect of Mars did not change the probability of Rasputin's death (there is no difference between prior and posterior probability). But as regards the poisoning, the probability of dying after the poisoning exceeded the probability of dying without the poisoning—while in the above analysis Strevens shows that the poisoning is not to be counted among the factors responsible for Rasputin's death!

The other criterion touches on counterfactual situations: if the present situation had not come about, the given event would not have happened. Concerning the effect of the gravitational pull of planet Mars on Rasputin's death, even if this was removed, Rasputin would still have died. Likewise regarding the poisoning: if the plotters had not poisoned the monk, he still would have been murdered. But, Strevens asks, what about the drowning? If they had not drowned Rasputin, would he have been murdered? And the answer is no. However, here Strevens gives an answer stemming from a counterfactual situation that does not accord with this answer and the above analysis: if the conspirators/murderers had not drowned Rasputin, they would have found an alternative way to do away with him (e.g., a bullet in the head).

In view of the analysis of the Rasputin case, the comparison of the Kairetic test for difference makers with these two criteria (probabilistic and counterfactual) awards victory to the former. This comparison is not the only interesting discussion that Strevens develops in his writings. It is noteworthy that he supported his approach by means of interesting analyses to which I cannot devote space here. I may allude to just one problem for which Strevens found a solution. From a logical viewpoint, if *B* stems from *A*, then *B* will also stem from the disjunction (*A or X*), where we may set in *X* whatever comes into our head; resorting to this logic, Strevens proposes a new disjunctive theory to explain Rasputin's death as the result of (being drowned *or* being poisoned). (In fact, disjunction can number many dozens of factors, X_1 or X_2, etc.) The solution that Strevens offers for the disjunction problem comes through the term *Cohesion*. According to this concept, if we look separately at the two components of the new theory (drowning and poisoning), then only one of them makes a difference regarding the monk's death. Thus the second factor weakens the cohesive value of the present theory, which therefore is superfluous.

The Kairetic model spurred several critical articles and discussions, which I shall not survey here (see, e.g., Levy, 2010; Potochnik, 2011; the debate between Strevens, 2008a, and Woodward, 2008). Still, I would like to conclude this short review with three comments.

First, the focal point of the present model seems to be an attempt to find in the network of possible causes those that make a difference to the given phenomenon. How is this done? By the elimination method, you have to rid the model you have in your hand—the primary model—of all factors (laws and descriptions of concrete situations) whose elimination will not affect the causal inference of the description of the given phenomenon. The immediate question is this: Why does the researcher believe that the list of causes contained in the primary model does indeed take into account the factor/s truly responsible for the occurrence of the given phenomenon? Assuming that in fact the primary model does not include the decisive factor, the eliminative procedure is liable only to impede the research process, because in this case the present procedure will only convince the researcher that he indeed has the correct model, based on the difference-making causal factor for the given phenomenon, when in fact this is not the case.

Second, it is not quite clear why one has to appeal to the Cohesion criterion to remove the possibility of creating a disjunctive theory. Is it not possible to get rid of the unwanted component by disjunction (drowning or poisoning) simply by applying the eliminative procedure? In other words, the disjunctive theory may be seen as a kind of primary model, from which may be removed the unwanted factor, namely, the death of the monk by poisoning, by the Kairetic procedure, so there is no need for the help of the Cohesion criterion.

Third, Strevens (2008b) suggests that in a complex system, the transitive quality of difference making is preserved: "If c is a difference-maker for d and d is a difference-maker for e, then there is a kernel for d containing c and a kernel for e containing d. These kernels may be put together to make a compound standalone explanation for e. The explanation includes c, so c is a difference-maker for e" (p. 122). Later Strevens goes on to say: "The difference-making relation is transitive (section 4.2) and causal claims assert relations of difference-making (this chapter). Therefore, *is a cause of* is transitive: if c is a cause of d and d is a cause of e, then c is a cause of e" (p. 208).

The question that arises is whether causality always shows transitivity. Let us examine the following examples. If we assume that the action of the handgun is an example that realizes a complex explanatory system of the kind under consideration (this system encompasses intermediary steps, mediation steps, in the chain of the explanation), then I am not certain that Strevens's claim for transitivity holds. Why not? For the following reason: we know that (a) squeezing the trigger causes (b) the explosion of powder in the cartridge, which causes (c) the firing of the bullet from the barrel of the handgun. Does this example substantiate causal transitivity? That is, can we suggest that when (a) is the difference-making cause of (b), and (b) is the difference-making cause of (c), (a) is also the difference-making factor of (c)? The answer, to my mind, is we cannot. Why not? Because many cases may be imagined where squeezing the trigger will not expel the bullet from the handgun's barrel (e.g., the firing pin is broken, the strike of the pin is very weak, the powder in the cartridge is damp, etc.). In fact, a considerable number of instances of this sort can be found. For example, causal transitivity is not preserved even in the following case: stepping on the gas pedal in the car speeds the action of the engine, which accelerates the car's movement. It is possible to visualize many cases where stepping on the gas will not speed up the car's movement. Here perhaps it is worth adding a clarification of what I mean. I believe that a handgun's system has no transitivity of the following kind: *A* is taller than *B*, *B* is taller than *C*, and therefore *A* is taller than *C*. The relation "taller than" is empirical, and measurable, so clearly the relation "*A* is taller than *C*" stands on its own account (*A* and *C* can be measured, and it is evident that this relation exists). Not so regarding the relation "squeezing the trigger causes the firing of the bullet." This relation is perhaps more similar to that in the following system: *A* hates *B*, and *B* hates *C*, but *A* does not hate *C*! (For brevity, I shall avoid discussing here the question when transitivity holds.)

OTHER MODELS FOR GIVING EXPLANATIONS

As stated above, the article by Hempel and Oppenheim of 1948 opened a special research field in the philosophy of science: study of the scientific explanation. Quite soon, it became clear that there was a need to build schemes of explanation for probabilistic/statistical phenomena (see "Statistical Models for Giving Explanations," above) and the concept of causality had to be viewed as extremely important in giving explanations (see "Causal Models for Giving Explanations," above). The two models that I review here, Kitcher's (1989) Unification model and van Fraassen's (1980) Pragmatic model, appeared even before the publication of Woodward's (2003) model and Strevens's (2008b) model; they were a source of influence and of controversy for these two philosophers. There is no subject that unites these two models, as there was in the two preceding sections (the subject of probability and statistics, and the subject of causality). The models address two different explanation topics: the Unification model centers on the idea that a scientific explanation deals with the ability of scientific research to understand a phenomenon through its inclusion in a unified theoretical/empirical scientific framework; the pragmatic model centers on the conception of explanation in the pragmatic framework and in the context of questions and answers.

The Unificationist Model

The basic notion behind models of Unification as the basis of scientific explanation arises from a study of the development of scientific knowledge. From this viewpoint it emerges that science advances by its unifying a large number of observations that at first glance seem devoid of any connection, by gathering them under the wings of a single scientific theory. The question, of course, is what the nature of the unification process is that serves as infrastructure for giving an explanation. The professional literature discusses two approaches to unification as the basis of explanation: that of Friedman (1974) and that of Kitcher (1981, 1989). Friedman tries to handle unification by means of laws or theories that succeed in explaining under given conditions a large collection of observations. (He encountered, among other things, the problem arising from the question, mentioned above, how is it possible to distinguish a true law from an accidental generalization, and in the present case a good unifying theory from a bad one? A further criticism, essentially technical, is summarized by Salmon, 1990.) By contrast, Kitcher seeks to base unification on the concept of explanatory patterns, which are used by researchers again and again to structure unified scientific knowledge. In the present survey I concentrate on Kitcher (1981, 1989) because his approach, which continues Friedman's ideas, is highly developed, and has attracted attention in the philosophical literature (e.g., it influenced Strevens's Kairetic model). Both approaches attempted to contend with the problem that arose already with Hempel and that is linked to the attempt to explain an empirical generalization by means of an extremely broad and deep empirical theory (e.g., explaining Galileo's law by Newtonian theory). The problem was technical/logical in essence and centered on the question, how is it possible to distinguish a broad/deep theory, whereby it is possible to explain the given generalization, from a theory which is not broad and deep, but can explain through deductive inference that generalization? (See discussion in Hempel, 1965; Kitcher, 1989; Psillos, 2002; Salmon, 1990.)

In contrast to Hempel's approach, and to the causal approaches that propose explanations for phenomena that can be assessed individually, Kitcher (1989) proposes to treat each and every explanation as part of a whole explanatory system. In his view, this unification is the way science develops; he writes:

> *Science advances our understanding of nature by showing us how to derive descriptions of many phenomena, using the same patterns of derivation again and again, and, in demonstrating this, it teaches us how to reduce the number of types of facts we have to accept as ultimate (or brute).* So the criterion of unification I shall try to articulate will be based on the idea that E(K) is a set of derivations that makes the best tradeoff between minimizing the number of patterns of derivation employed and maximizing the number of conclusions generated. (p. 432, italics in original)

(To clarify E[K], Kitcher writes: "E(K) is to be the set of derivations that best systematizes K . . . is the set of derivations that best unifies K" [p. 431].)

The classic example, it may be said, which substantiates the Unificationist model, is Newtonian theory, which took under its wing diverse fields of research, for example, movement and free fall of bodies to the earth, movement of stars in the solar system (including Kepler's laws), and high tide and low tide according to the pull of the moon. Unification of all these phenomena in the framework of Newtonian theory was accomplished, in Kitcher's opinion, by repetitious use of the same explanatory patterns: for example, use of Newton's laws and their application to different cases by means of explanatory patterns of differential and integral calculus. In this case the unifying power is especially high, because with a very small number of explanatory patterns, good explanations were created for a large collection of observations.

Clearly, the Unificationist model is anchored to deductive explanatory patterns. In this regard the D-N model and the Unificationist model are alike: both use logical and mathematical languages. The difference lies in Hempel's explanatory model being local—that is, the explanation applies separately to each and every observation, while Kitcher's Unificationist model is holistic. In the present model, the explained observation is an integral part of the same scientific knowledge that has been integrated by the same explanatory pattern—namely, the same explanatory pattern that has been applied to the given observation also dealt with very many other observations.

Here the following immediate question arises: Can Kitcher's Unificationist approach contend with the problems raised against Hempel's D-N model? Kitcher answers in the affirmative. Let us peruse again the illustration of prediction of the height of the flagpole from the length of the shadow. Kitcher's solution to this problem rests on the following argument: since the concept of explanation is anchored to the concept of unification, then explanations based on inferences from the premises to the conclusions, that is, from objective qualities (such as the flagpole's size) to their effect (such as the shadow's length), are more numerous and united than explanations obtained from inferences in the reverse direction (from conclusions to premises)—that is, from the shadow to the properties of the object. This argument is especially reinforced if we consider that one may learn about objective properties even if they are not lit up (i.e., they do not cast a shadow), or they are wholly transparent or are so large, like a galaxy, that no shadow is to be found from which we might learn something about their size, the galaxy's size. In other words, the explanation for the size of the flagpole based on the shadow's length is lesser than an explanation of the length of the shadow cast by the flagpole, because the latter is well woven into a unified system of knowledge, while the former is not. In other words, the first explanatory pattern (from premises to conclusions, from the flagpole's height to the shadow's length) is applicable to and acceptable by the explanatory system prevalent in science, while the other explanation (from conclusions to premises, from the shadow's length to the flagpole's height) is not so applicable. The addition of the latter to the group of explanatory patterns accepted in the research field does not enrich the conclusions domain like the first explanatory pattern, but unnecessarily enlarges the number of explanatory patterns, and thus lowers the value of scientific unification. This is Kitcher's way of perceiving the notion of causality (and the explanatory asymmetry in the example of the flagpole and the shadow) through the notion of Unification.

The Unificationist model generated much criticism. Let us look at three instances. The first contends that the basis of the concept of Unification should be the concept of explanation, and not the reverse, that is, unification as the infrastructure of the explanation. For example, Unification, which Newtonian theory irradiates onto a large collection of observations, is grounded in the theory's ability to explain under certain conditions a large array of observations (i.e., it originates in the theory's ability to answer many *why/how* questions); by comparison, it grates somewhat to hear the contrary argument: Unification, which was created by the same explanatory patterns, is responsible for the explanation of the collection of observations (that is, Unification is responsible for answers to *why/how* questions).

The next criticism raises the possibility that there is large number of unification processes that do not furnish a basis for explanation, and the present model does not offer a clear-cut distinction between a Unification process that is explanatory and one that is not. For example, a catalogue that details all the flowers and all the trees does not provide an explanation for the distinction between a flower and a tree, but only an answer to the question if this particular item is classifiable as a flower or a tree.

The third criticism casts doubt on the idea that the explanatory pattern is the sole source of scientific unification. Let us look at the research process in psychology. I shall not miss the mark if I say that all research studies in psychology rest on the use of inferential statistics, for example, use of t tests or ANOVAs to obtain the research hypothesis (H_1) and to reject the zero hypothesis (H_0). Has the use of these explanatory patterns time after time over decades in psychology designed psychology as a unified science? The answer is a decisive no. Unification is not the foundation of vision, perception, personality, social and cognitive psychology, and other such diverse research domains in psychology. There is no room here for a comparison to the unity (Newtonian) that appears in classical physics. Furthermore, even though sociology and criminology use inferential statistics, the degree of scientific unification of these scientific disciplines is extremely small. What does this comparison of the natural sciences with the social sciences say? To my mind it points out that explanatory patterns are not the decisive factors that make a certain research field unified. There are other factors here. For example, Rakover (2012b) suggested that regarding the present discussion the difference between psychology and physics lies in the theory and practice of measuring the studied behavior—in psychology measurement units as in physics have not yet been discovered (e.g., units of measurement of length, weight, and time).

The Pragmatic Model

In his book *The Scientific Image*, van Fraassen (1980) proposes a new model for giving explanations that centers on two key points. First, van Fraassen views explanation as giving an answer to the *why* question. Secondly, as an explanation, as an answer to the *why* question, he emphasizes the pragmatic aspect in this process, that is, the fact that the question and answer are context-dependent. It is essential to know in what context the question was raised and answered. To clarify this issue, let us look, for example, at the following routine question: Why did David walk to the cinema?

According to the context, the question may focus on three viewpoints in this sentence:

1. Why did *David* walk to the cinema, instead of, for example, Johnny, Ruthi, and so on?
2. Why did David *walk* to the cinema, instead of going by car or bike, running, flying, and so on?
3. Why did David walk to the *cinema*, instead of a café, the sea, the pool, watching TV, and so on?

For each context, answers may be proposed with different emphases. For example, for (1) the answer might be because only David got a free ticket from a friend; for (2) it may be because David lives very near the cinema; for (3) it may be because David is fed up with cafés, the sea, and the pool, or because he hasn't been to the cinema for over a year.

In light of context considerations of this kind, van Fraassen offers a formal wording of the *why* questions and their answers. (It is worth noting that this approach of van Fraassen's was based on a certain kind of logic, erotetic logic, which concerns formal analysis of presentation of questions and giving answers.) In general, the question and answer may be formulated thus:

Why did phenomenon *P* occur?
Because of cause (*C*).

He proposed that questions/answers of this kind are based on the following three fundamental concepts: P_k, *X*, and *R*, where

a. P_k denotes the subject of the question, what is to be explained, the phenomenon under discussion. For example, with the above example, the subject of the question is walking to the cinema, that is, why David walked to the cinema.
b. *X* denotes the contrast class, alternative subjects. In our example it may encompass various people, various ways of action, and various places.
c. *R* denotes a relevance relation in the presentation of the question and in giving the direct answer to it. In the above example the answer says: compared with all possibilities of *X*, David walked to the cinema because *C*.

The explanation, the answer, favors or supports what was presented in the question, and not the possibilities that appear in the contrast classes. Usually the cause is linked to a description of a series of events that leads to the occurrence of the given phenomenon.

On the importance of the contrast-class and the relevance relation, van Fraassen writes:

> The description of some account as an explanation of a given fact or event, is incomplete. It can only be an explanation with respect to a certain *relevance relation* and a certain *contrast-class*. These are contextual factors, in that they are determined neither by the totality of accepted scientific theories, nor by the event or fact for which an explanation is requested. (1980, p. 130)

Van Fraassen holds that the earlier approaches to explanation erred in that they did not take into account the context. He writes:

> The discussion of explanation went wrong at the very beginning when explanation was conceived of as a relationship like description: a relation between theory and fact. Really it is a three-term relation: a relation between theory, fact, and context. . . . Being an explanation is essentially relative, for an explanation is an *answer*. . . . But exactly what is requested, by means of the interrogative "Why is it the case that P?", differs from context to context. (1980, p. 156)

In his view, contextual factors can explain problems that other models struggled with, frequently without success. He writes the following on the problem of the asymmetry that finds expression in the flagpole and shadow illustration (see above):

> If the asymmetries of explanation result from a contextually determined relation of relevance, then it must be the case these asymmetries can at least sometimes be reversed by a change in context. In addition, it should also be possible to account for specific asymmetries in terms of the interests of questioner and audience that determine this relevance. (1980, p. 130)

And certainly, to substantiate this argument, van Fraassen tells a story he heard at the place where it happened and in the presence of its protagonists. It turns out that a man, who was a friend of van Fraassen's father, fell madly in love with a young servant girl and expressed his love for her on the balcony of his house. This man killed the beautiful maid out of unconquerable jealousy, and built a tower at the place where the girl was done to death. The tower was built to a certain height, so that the moment the sun began to set, the tower would cast a shadow over the balcony of the declaration of love, and plunge it into darkness. So here before us is an account of events that explains why the tower was built to certain height with the aid of the shadow's length. What happened here is that the change in context, and in the knowledge of the tower builder's intention, could explain the tower's height by means of the shadow's length.

Salmon (1990), who reviewed van Fraassen's pragmatic approach to explanation, and quite naturally discussed the story of the tower and the shadow, is unwilling to accept this narrational explanation. Here is what he writes about the concept of relevance relation:

> And the relation R is a relation of *intentional relevance*. Our remonstrance was based on the belief that this relation is not a suitable explanatory relevance relation. As an advocate of causal/mechanical conception of scientific explanation, I am not prepared to admit that *effects* explain *causes*, even where *conscious purposeful* behavior is involved. (p. 144; my italics)

Here I must make the following comment. On the one hand, it is understandable why Salmon, who is committed to the physical explanatory model, the C-M model, discussed above, finds it hard to accept explanations based on subjective concepts such as intention, goals, and consciousness; but on the other hand, it seems to me that one cannot explain human behavior, or the behavior of developed animals such as primates, dogs, and cats, without using these concepts (in a later chapter I shall deal with these subjects). To support this latter statement, I shall describe a relatively simple episode concerning the behavior of alley cats.

On cold winter days I noticed that when I parked my car in a covered parking space under my apartment building, two or three cats immediately gathered near the vehicle. They stayed there, waiting for the moment I disappeared into the corridor leading to the elevator. One day I was overcome with curiosity, so I tiptoed back to see why the cats had assembled around my car. What I discovered was three cats curled up snoozing on the hood of the engine. *Aha*, I said to myself, *so that's the reason*: they understand that a car that drives into the parking space has a warm and cozy engine hood. And why do they wait till I leave? So that I won't chase them away. And sure enough, one of the cats looked up and stared at me, all tense. But I felt that they were so smart that they deserved to enjoy the warmth, despite the splashes of mud they spread on the windshield and the hood. Apart from that, they were so handsome. (Of course, one may present a more behaviorist explanation, and not so mentalist. But this is not the place to debate the matter; that discussion will unfold in the following chapters.)

SUMMARY

Table 2.1 summarizes the nine models surveyed above. It contains the model's name; the model's basic structure; the model's application to events, generalizations, or both; and one important problem that the model finds difficult to handle.

In this chapter I reviewed nine different models that, in my opinion, are likely to be of interest to psychology students (the philosophical literature deals with an additional number of models that I have not discussed here; see, e.g., Salmon, 1990). All the reviewed models here attempt in their own way to offer an answer in the framework of the rules of the game of science to the *why* question (there are also several models that try to answer the *how* question, e.g., the C-M model). As may be seen from Table 2.1 the models offer different answers to the *why* question, and thus they express different conceptions of the structure of the scientific explanation: from the approach that a scientific explanation is based on a deductive argument (the D-N model) to the approach that sees in explanation an answer to the *why*

Table 2.1 The Nine Models

Model	*Name*	*Model Structure*	*Application*	*One Problem*
1	D-N	Deductive argument	Mainly Events	Flagpole
2	D-S	Deductive argument	Generalization	Accidental generalizations
3	I-S	Inductive argument	Events	Low probability
4	S-R	Prior/posterior probabilities	Events	Causality, experimentation
5	C-M	Causal process	Events	Causal relevancy
6	M-I	Causality/ counterfactuals	Events	Noncausal accounts
7	Kairetic	Causal difference making	Events	Incorrect factors
8	Unificationist	Explanatory patterns	Both kinds	Unification is not a basis
9	Pragmatic	Contextual factors	Events	Causal/mechanical process

question that varies according to the contextual factors and the interests of the explainer and the questioner (the pragmatic model). Considering this, it may be said that the conclusion that arises from Table 2.1 is this: although all these models work in the accepted framework of the methodology of science, *there is no accepted uniform conception of a scientific explanation.* This conclusion stands in contrast to the wish of Hempel (1965), father of philosophical research into scientific explanation, who proposed seeing his models as an all-inclusive explanatory framework:

> The central theme of this essay has been, briefly, that all scientific explanation involves, explicitly or by implication, a subsumption of its subject matter under general regularities; that it seeks to provide systematic understanding of empirical phenomena by showing that they fit into a nomic nexus. (p. 488)

Chapter 3

Models of Explanation (II)

Natural and Social Sciences

As I pointed out before, Hempel regarded his models as a general format for giving explanations in various research fields, for example, the natural sciences and history. The nine models reviewed in chapter 2 were tailored to research fields in the sciences (note that Salmon developed the C-M model as an explanation of phenomena in the natural sciences except quantum mechanics). I shall not be far wrong in saying that these models are not readily applicable to the fields of social sciences and especially psychology. To clarify this matter, let us examine the following illustrations.

a. David moved his knight two squares forward and one square left, and declared a victory, checkmate; David stopped his car before the intersection when the light changed from green to red.
b. The heart's task (function) is to pump blood through the circulatory system.
c. David traveled from Haifa to the Tel Aviv Opera House to hear the diva sing Puccini's *Turandot*; David climbed up a ladder to take a Hebrew–English dictionary off the top shelf.

I believe that if we attempt to match each of the aforementioned nine models to each of these examples, we shall encounter many difficulties. For example, according to the D-N model (and according to Hempel's two statistical models), to explain these observations, we must find a law or a well-grounded empirical theory whereby we will be able to predict these observations under certain conditions. But it is easily seen that there are no laws or empirical theories relevant to these kinds of behavior: there are no natural laws for the game of chess; there is no law that determines that one must travel to Tel Aviv to hear an opera; there is no law for climbing a ladder; and I believe I am not mistaken in saying that there is no law of nature that assigns to the natural heart a function of activating blood circulation. (Why? Because the blood may be circulated by means of an artificial heart or a machine.)

The behavior described in (a) is characteristic of people who have resolved to uphold certain arbitrary rules for various reasons. In chess, the knight (horse) cannot be moved right or left, forward or backward, or over three aligned squares, where the knight leapfrogs two squares and lands on the third, which continues the preceding two in a straight line. That might make for an interesting game, but it is not the regular game of chess. And whoever runs a red light is liable to cause a serious road accident or be made to pay a heavy fine. Since all the foregoing nine models seem made to measure for research fields in the natural sciences (physics, chemistry), it is hard to see how they can contend with the following two facts: people behave in keeping with game rules formulated and legislated by people; they obey these rules, in the end, out of their own free will.

The observations in (c) describe teleological behavior. People perform actions in order to realize certain purposes (to hear an opera singer, to peruse an English–Hebrew dictionary). And again it is hard to see how the nine above models can handle an explanation of teleological behavior, where a future purpose determines present behavior. This situation is not simple for models that one way or another postulate causality from past to present (or from present to future)—that is, event *E*, which precedes in time event *G*, explains the occurrence of *G*, but not the reverse. In observations of the kind (c), the direction of the causality can be reversed. For example, it is reasonable to suppose that the future event, the goal *G*, is represented in a person's cognitive system as *G**, so one may say that because *G** precedes in time action *A*, *G** may explain causally the occurrence of *A* (e.g., David thought about the diva who was to sing an opera in the future, *G**, and as a result decided to travel to Tel Aviv). However, although this solution seems to settle the timing problem, such that the goal represented in the cognitive system has become a cause of the action, in fact this solution raises a known and unsolvable problem. It is hard to explain, to suggest a solution to, the following question: How did a thought about the diva in the opera at Tel Aviv cause a real action? We can explain how physical event *A* causes the occurrence of physical event *B*, but we still have no notion how a mental event causes the occurrence of a physical event, and the reverse (see Robb & Heil, 2014; Yoo, 2015). This is a complex problem, linked to the great unanswered question of the connection between consciousness and the brain, a problem for which no theory has been found that explains the amazing yet routine connection between mind and body—that is, we still do not possess a mind/body theory (see Rakover, 1990, 2007, 2011/2012a, 2011/2012b).

The observation in (b) speaks to teleological behavior (the goal/function of the heart is to operate blood circulation). But unlike observation (c) (to hear an opera singer, to peruse an English–Hebrew dictionary), which involves processes of consciousness and representation of events in the cognitive system, in the case of the heart (and, as a further example, of the kidneys, whose goal/function is to cleanse the blood of its accumulated toxins), it is hard to suggest that these systems are linked to consciousness. Philosophers of science have tried to base a functional explanation on a causal explanation, and I shall review this effort briefly later on.

I shall attempt to deal with part of these questions in the present chapter. In the following chapters I endeavor to propose a methodological solution for phenomena connected to the behavior of animals, more particularly humans, behaviors intertwined with behavioral/neurophysiological components and at different levels of consciousness and awareness. In this chapter I highlight a number of problems hinting that direct application of the nine foregoing models to research fields in the social sciences is extremely difficult (if not impossible); therefore, models of explanation intended for these fields need to be developed. First I briefly review explanations by means of rules, then functional explanations, and finally I concentrate on the modern approach of development of an explanation by means of a mechanism, and the application of this model to cognitive psychology.

THE RULE-EXPLANATIONS MODEL

We shall begin the discussion of rule-explanations with a simple illustration (observation [a] above): David braked his car before entering the intersection when the traffic light changed to red. How could this behavior of David's be understood, namely, his stopping the car? This behavior is very hard (if not impossible) to explain by any of the nine models reviewed in the last chapter. For example, any of Hempel's three models requires reference to a law of nature, a well-founded theory or a stable empirical generalization (namely, a generalization that has acquired broad empirical anchorage). However, as can be understood immediately, and as I pointed out above, no natural law, no well-founded theory, and no stable empirical generalization exist by whose means David's behavior can be explained as a private case of these theoretical/empirical structures. Why? Simply because David's behavior is no other than a kind of obeying or behaving according to the road rule that states that he must bring the car to an absolutely stationary state before the intersection when the traffic light changes to red (for simplicity, I call this the "red-intersection rule"). This road rule instructs the driver how to behave (how to drive the vehicle) when he reaches a traffic-light-controlled intersection. This is a behavioral rule that requires the driver to do what he is told. It is not like a law of nature that has been revealed in research, it is not a theory about behavior of certain objects, and it is not a kind of empirical generalization discovered by statistical tools (which have matched to an array of observations a certain mathematical function). This rule is not like any of these. This is a rule that people invented so as to order the movement of traffic at the entrance to an intersection.

So how shall we explain David's behavior by means of a road rule? The explanation is this: David, who learned the traffic rules used in the state of Israel, followed them as required when he drove his vehicle toward a traffic-light-controlled intersection; that is, when he discerned that the light at the intersection was red, he stopped his car. In the same way we may say that David played chess (see example [a] above) when he moved the pawns according to the rules of the game, which he had learned and practiced before. As seen from these two examples, a large part of human behavior is explicable through appeal to the rules of behavior (including social norms) that people have invented so as to order the relations between person and person, and between person and environment (e.g., after the picnic, kindly deposit the trash in the bin).

A large portion of the rules of behavior can be regarded as possessing the *if-then* scheme (e.g., Reese, 1989). This view makes it possible to give rule-explanations the following form:

1. Traffic rule (red-intersection rule): If the traffic light at the intersection is red, then the vehicle has to be brought to a halt before entering the intersection.
2. Practical conditions: (I) The traffic light at intersection (a) has changed to red; (II) David approached intersection (a) in his car; (III) David is a good citizen who obeys the traffic rules.
3. Conclusion: David will stop his car before intersection (a) as required.

And because David indeed stopped his car before intersection (a), we shall suggest that David's behavior obtained an explanation by appeal to the traffic rules (that is, with the aid of the rule-explanations model). If David crossed the intersection by running a red light—namely, he violated the foregoing traffic rule—we would have to look for another explanation that answers the question of why David disobeyed the traffic rule. (Maybe he was in a hurry to an important meeting, maybe he was escaping someone, and the like.)

As may be seen from the present explanation, several differences exist between rules of behavior and laws of nature, theories, and statistical generalizations (e.g., Carroll, 2010; Rakover, 1990; Reese, 1989). I shall concentrate on seven important properties that distinguish the concept "law of nature" from "rules of behavior," which were created by human beings (properties likely to depend on each other in certain cases):

1. *Universality*: While natural laws (determinist or statistical) are universal (e.g., the general theory of relativity applies not only to the solar system, but also to the Milky Way and other galaxies), rules of behavior are limited and depend on context, society, and social groups (e.g., in Israel, one does not turn right at a red light; in several U.S. states, one may turn right if there is no oncoming traffic and the intersection is empty).
2. *Exceptions*: While natural phenomena do not depart from the description that appears in natural laws (even if the law is statistical), human beings can break rules for various reasons and pretexts. (For example, the stone cannot stop falling to the ground, but the person can disobey the traffic rule out of her own free will, wittingly, and run the red light. David in the above example did not act in that way, because he is an obedient type.)
3. *Consciousness*: While natural phenomena occur mechanically, do not depend on conscious guidance, and behave according to the description given by the natural laws, rules (such as road rules) have no meaning except in the human being's conscious system. (If David had been unaware of the road rules because he had not learned them or had forgotten them entirely, he would have crossed the intersection against the red light; but the stone had been falling to the ground for millions of years even before Galileo's law was revealed.)
4. *Arbitrary changes*: While a natural law is not subject to an arbitrary change (because the change requires an enormous theoretical and empirical effort, and a random change would be rejected at once), rules formulated by the human being are subject to arbitrary change. A law is the outcome of a research discovery that describes how nature behaves, while a rule is a human being's invention meant to establish a new behavioral order (to order behavior or to amuse the human being). (For example, it can be decided that traffic lights will have only two colors instead of three: green [go] and red [stop], and the amber light in between will be dropped; the length of time each color is on can be increased or decreased arbitrarily; it can be decided that the knight in the chess game will not leapfrog the opponent's pieces; and women may be required to open the door for men, and not the opposite, as is nowadays accepted. But it is not possible to change the power in Galileo's law from 2 to 3.25, or to decide to multiply the energy variable in Einstein's famous law by some constant [e.g., 0.69].)
5. *Estimation*: While natural laws are estimated essentially by an empirical test (a match of predictions to observations), rules (such as road rules) are estimated by the measure of their suiting the purpose for which they were created (e.g., road rules were created to facilitate the flow of traffic).

6. *Causality*: While the explanation for the given phenomenon rests on a causal natural law that determines that the event occurred for the reason described in the law, rules do not constitute reasons that dictate behavior but direct and guide how one should behave. For example, the driver's knowing the traffic-light rules does not dictate her obeying them. True, she can be penalized, or she may cause a serious accident, but the behavior is not performed automatically and mechanically as in the case of the natural law. Reese (1989) writes on this: "Knowing a rule in the first sense (being able to state it) cannot by itself cause behavior; it can be at most only a normative disposition to behave" (p. 34).
7. *The subject of discussion*: While laws of nature concern changes of energy and properties of materials and their changes, rules of behavior concern information, namely, representations that appear in the individual's cognitive system: for example, the passage from potential energy to kinetic energy, from kinetic energy to heat energy. By contrast, the traffic-light rule constitutes information stored in the individual's memory and directs him how to behave in the vicinity of an intersection where the traffic light shows red, meaning stop.

Although we have noted that consciousness is one of the properties distinguishing natural laws from rules of behavior, we should emphasize here that cognitive psychology speaks of certain kinds of rules that the individual is not aware of, and that apparently are part of the human being's genetic equipment. The intention here is first and foremost phenomena connected to the acquisition of language, to behavior of attention, perception, memory, and so on (see Chomsky, 1980; Pylyshyn, 1980, 1984). Behaviors of this kind are explained by appeal to symbolic representations in the cognitive system to which are applied certain rules similar to the computational operations that the computer performs on the physical representations located in its system. For example, in every person there is a very basic system of rules of grammar, through whose automatic direction and guidance people are able to produce, each in his or her own language, properly constructed sentences. Thus, Reese (1989), citing Searle (and adding that Chomsky agreed with Searle about this), writes: "The rules of language are not like the laws of physics, for the rules must do more than describe what happens, they must play a role in guiding behavior" (p. 37).

These unconscious rules are different from conscious rules (e.g., they do not uphold properties [3] consciousness or [4] arbitrary changes), and hence they gradually approach natural laws. Nevertheless, and although these unconscious rules are apparently part of the person's genetic equipment, they are different from natural laws especially in the following properties: they are not universal, like the laws of physics (see [1]); they are information processing (see [7]); and they direct and guide behavior (see [6]). Toulmin (1974) also considered this distinction between conscious and unconscious rules: "At one extreme, we explain the things that befall human beings in terms of 'rules' without implying that the generation of these events involved any element of rationality of consciousness. At the other extreme, by contrast, the notion of rules is linked quite rigorously to 'conscious' employment of 'rational' procedures" (p. 190).

So I have apparently set forth quite a few properties that distinguish rules from laws that support the argument that a rule-explanations model cannot be conceptualized in the framework of the models proposed for the sciences, namely, the models reviewed in chapter 2. Does it follow that the rule-explanations model is unique?

To support an affirmative answer with a proposition on the specificity of rule-explanations, I must show in addition to the foregoing that this kind of explanation cannot be taken as a special case of the teleological (on which I shall enlarge later on). Why? Simply because one may argue that the rules of behavior were created by human beings in order to serve a certain purpose (as stated, to organize the flow of traffic, people's behavior, and so on). For example, the explanation of stopping the car at the red light can be attributed a teleological character in this way:

> David stopped his car before entering the intersection at the red light *with the goal of* realizing the red-intersection rule as a good citizen who obeys the law.

From this explanation it emerges that David stopped his car in order to attain certain ends, when the means, the way to attain these ends, was paved by the red-intersection rule. Despite the temptation to grasp rule-explanations as a kind of teleological explanation, I shall enumerate here several properties that separate these two kinds of explanation, and that support the conception of rule-explanations as specific and unique.

1. *It is difficult to characterize all behaviors by means of rules*: For example, prejudice in a person of race A against a person of race B (such as a white person's feeling of superiority over a black person) is an emotional/cognitive response that arises in A upon catching sight of a person of race B. Also, avoidance of a place that is liable to cause one harm and seeking food some five hours or more after the last meal—all these and similar cases are easy to characterize as responses to a certain kind of stimuli. These are examples of stimulus-response connections created as a result of learning or of an innate tendency.
2. *People are not always aware of the purpose of behavioral rules*: We all perform much of our routine behavior according to social rules we have

learned by imitation, without being aware of the reason why these rules were created. This is especially evident in respect of religious rituals of ancient tradition not understood by the modern person, or with regard to rules with a lengthy history. For example, why does one wear a tie around one's neck? Today the answer rests on reasons of fashion. But history teaches us that in the Thirty Years' War that raged in Europe, mercenaries tied a scarf around the neck to wipe the mouth while eating so as to protect the shirt from stains. The French word for this accessory—*cravate* (tie)—developed from the name of the mercenaries, Croats. Because we are not aware of the unconscious cognitive rules that I discussed above (about learning a language, perception, and the like), clearly, we are unaware of their purpose.

3. *For the same goal there are different rules, and for the same rule there are different goals*: For example, a green traffic light is a direction to the driver to advance in the permitted driving lanes, where the goal is to maintain an orderly and controlled traffic flow. However, this does not hold for David and Dan, who together have agreed on a new game rule, where a green light is a signal to start a race for who is the faster driver. And the reverse: obeying the red-intersection rule is for Dan a way to Ruth's heart, because his goal in obeying this rule is to impress Ruth that he indeed is a responsible, dependable person!
4. *Even if theory* T *explains behavior rule* A, *it is not always possible to explain by means of* T *behavior* X, *which is explained by* A: For example, it is hard to see how *T*, a learning theory that explains how people have acquired the red-intersection rule, is likely to explain the observational fact that David stopped his car before the intersection when the traffic light changed to red. The explanation of the learning theory for this behavior is this: the color red is a discriminative stimulus that arouses in the driver the braking response. This connection between the discriminative stimulus and the braking response is acquired by giving many reinforcements. This theory will be hard-pressed to explain this fact: David obtained his driving license after reading only once the traffic rules, and passed the theory and practical driving test at the very first attempt. That is, the explanation for David's correct behavior will be based on activation of the knowledge of the red-intersection rule and not on the explanation based on theory *T*, since this theory requires a long period of training to establish a stimulus-response connection.

In sum then, these properties apparently succeed in distinguishing a teleological explanation from rule-explanations, and it does not seem possible that the former explanation will be able to embrace the latter, and the reverse. Each of these two explanation models (rule-explanations, teleological explanation) stands alone, and various behavioral phenomena are conveniently explained by their means.

FUNCTIONAL, TELEOLOGICAL MODELS OF EXPLANATION

In 1942 Hempel wrote a paper proposing that explanations in history should be given in the form of the D-N model, which he was to publish later, in 1948 and 1965. In his opinion, it had to be shown that the historical event under study could be predicted to occur given several conditions and general laws. Hempel came out against the prevalent opinions in the history sciences before then. Previously, the prevalent view, enunciated by the German philosopher Wilhelm Windelband (see the review by Psillos, 2007), was that the sciences and the history fields had to be sharply distinguished methodologically: the approach to the former was to be *nomothetic*, to the latter *idiographic*. The former approach aimed to reveal the laws of nature (through a methodology using experimental and control groups) and through them to understand the given phenomenon; by contrast, the idiographic approach sought to understand the individual's behavior, and to postulate value-judgments whereby the question why a studied historical event was of special importance would be asked. In German philosophy this distinction is linked to the important distinction that assigns to the natural sciences a methodology aiming to arrive at an explanation (*erklären*) of the given phenomenon, while in the history sciences one seeks to achieve understanding (*verstehen*) of the historical event. From this starting point, the German philosopher Wilhelm Dilthey developed the approach to understanding a historical event by the method of empathy: creating anew in the researcher's imagination the environmental and mental state of the historical figure as a way to explain his/her behavior in the framework of the development of historical events.

The Canadian philosopher William H. Dray opposed Hempel's approach to explanation (see Dray, 1957). He proposed that historical explanation is not causal (nor embraced by a law or well-grounded theory, as Hempel maintained), but is rational, that is, based on a reason. We explain the action of a historical figure by pointing out that in the given situation that figure's activity was rational (e.g., *X*'s behavior was logical and necessary in the reality surrounding *X*). This approach was criticized by Davidson (1963), who suggested that reasons were probably causes. In his view, Dray's "reason" was a kind of intentional explanation of action (e.g., *X* acted in a certain way that would satisfy his will), and it was nothing but a causal explanation even though it was not necessarily based on a natural law. (See discussion of Davidson's approach in Rakover, 2007; Robb & Heil, 2014; Yoo, 2015.)

Intentional explanations seem like teleological explanations, where event *A* occurred so (in order) that event *B* would occur. The question arising here, the "specificity" question, is this: Is the teleological explanation a one-of-a-kind, specific explanation, or is it nothing but a particular expression of a causal explanation? The idea underlying this question arises from Darwin's theory of evolution. This showed how it was possible to change a teleological explanation, such that future event *B* constitutes a cause of the appearance of event *A* in the present, by use of selection mechanisms and environmental adaptation that have occurred in the past and explain various developments in living creatures.

To discuss this question, I distinguish (in keeping with the foregoing) two kinds of teleological explanation: *functional explanation* and *will/belief explanation*. The fundamental differences between these two kinds are these: in the will/belief explanation the goal (the behavior's purpose) becomes an event that precedes the behavior under study itself by representation of the goal in the appropriate system (e.g., the cognitive system in humans and supreme animals), not so in functional explanations (e.g., the action of blood circulation as the goal of the functioning of the heart is not represented in an appropriate system and hence becomes the cause of the creation of the heart); the will/belief explanation occurs in the human's cognitive/consciousness system (and reasonably, partially in the cognitive/consciousness system of supreme animals such as primates and dolphins), not so the functional explanation, which is not required to assume consciousness and awareness. Similarly, Nagel (1961) says this:

> Indeed, it seems safe to suppose that biologists would generally deny they are postulating any conscious or implicit-in-view even when they employ such words as "purpose" in their functional analyses—as when the "purpose" (i.e., the function) of kidneys in the pig is said to be that of eliminating various waste products from the bloodstream of the organism. (pp. 402–403)

I shall deal with the will/belief explanation in later chapters, arguing that this explanation constitutes a procedure (scheme, model) that is one of a kind. In the present chapter I briefly review the discussion of specificity in the context of the functional explanation.

The literature discusses two main approaches (answers) to the specificity of functional explanations. Both attempt to show that the functional explanation is not a kind of special explanatory model. The first approach seeks to show that the functional explanation is anchored to a causal explanation that answers the question of *why* a certain element possessing a certain function appears in a given biological system. The answer generally proposes that this element is vital for the system's survival (i.e., the answer is anchored to the theory of evolution); the second approach seeks to show that the functional explanation apparently answers the question of *how* a certain element succeeds in contributing causally to the general functioning of the system in which this element appears (i.e., the answer proposes that the component has an important causal function in the system's functioning).

Why Does a Particular Element Appear in a Given System?

In chapter 12 of his well-known book, Nagel (1961) analyzes the teleological explanation and shows that it can be exchanged for a nonteleological, causal explanation that is identical to the teleological explanation. We shall look at the following example of the pig's kidneys (for simplicity, I eliminate from this example some variables such as the system's layout and environmental and internal conditions).

Teleological Explanation

The purpose (i.e., the function) of kidneys (*A*) in the pig (system *S*) is to engage in eliminating various waste products from the bloodstream of the organism (process *P*). Or in an abbreviated symbolic form:

> The purpose (i.e., the function) of (*A*) in (system *S*) is to engage in *P*.

And now Nagel suggests translating this teleological explanation into a nonteleological (causal) explanation that shows the vital necessity of *A* (the kidneys in the pig):

Nonteleological Explanation

a. Every system *S* (e.g., pigs) engages in *P*.
b. If *S* does not have *A*, then *S* does not engage in *P*.

Conclusion: *S* must have *A*.

And here is one more example, concerning the teleological explanation for the appearance of chlorophyll in plants, which Nagel (1961) uses several times:

> *The teleological explanation*: "The function of chlorophyll in plants is to enable them to perform photosynthesis" (p. 403).
>
> *The causal explanation*: "Plants perform photosynthesis only if they contain chlorophyll," or in an alternative formulation, "a necessary condition for the occurrence of photosynthesis in plants is the presence of chlorophyll" (p. 405).

Nagel's approach, whereby the teleological explanation may be expressed as a causal explanation without translation from one form of explanation to another causing any flaw, encountered much criticism, some of which Nagel parried in his book. I shall skip this discussion and just mention one critique that seems to me extremely severe (and, I may add, my impression is that Nagel's response to this criticism did not hit the mark). We may examine this criticism with this example: the function of the heart is to deliver blood to the circulatory system (which is vital for survival). The wording of this functional explanation in the causal form is: a necessary condition for the blood to flow in the circulatory system is the action of the heart (therefore, the heart appears in an animal that possesses a circulatory system). The criticism against this causal explanation holds that it is mistaken: the heart is not a necessary condition, because it can be replaced by an instrument, a special machine, which will perform the function of pumping blood through the system, or because it is possible to implant an artificial heart instead of the natural heart. In light of these cases, it cannot be argued that the presence of the natural heart is necessary for blood to be pumped through the circulatory system.

How Does a Certain Element Succeed in Contributing Causally to the General Functioning of the System in Which It Appears?

In 1975, Cummins proposed a different interpretation of functional explanations (see also Cummins, 1983: in this book Cummins applies his approach to psychology). In his view, the functional explanation does not clarify *why* a certain element appears in a given system (due to evolutionary development), but *how* this element contributes to a system's ability as a whole to function properly. That is, Cummins's functional analysis shows how the functioning of the system in its entirety is based on the functioning of the system's elements and the way these elements are organized so that the system as a whole behaves as it does. (The functioning of each of the parts of the system may also be decomposed into simpler parts, a breakdown that repeats itself recursively to the level where the process can be halted because the product of the decomposition is so simple that there is no point in yet another one.) Let us look once more at the functioning of the heart. Research on the circulatory system will distinguish at least four elementary components: heart, arteries (through which the blood flows *from* the heart), veins (through which the blood flows *to* the heart), and fluid flowing through the system, namely, blood. This will lead the researchers to see the heart as an instrument that functions as a pump that propels the blood through the circulatory system. This is an example of functional analysis substantiating the likelihood that this analysis will suit a wide array of research subjects and fields.

In his book *The Nature of the Psychological Explanation*, Cummins (1983) proposes an explanation for a complicated system by conducting a functional analysis, such as an explanation of the operation of assembly-line production:

> Production is broken down into a number of distinct and relatively simple (unskilled) tasks. The line has the capacity to produce the product in virtue of the fact that the units on the line have the capacity to perform one or more of these tasks, and in virtue of the fact that when tasks are performed in a certain organized way—according to a certain program—the finished product results. (p. 28)

Diagrams in electronics, computer programs, physiological analysis of an organism, are examples of this kind of functional analysis. This kind of analysis, then, breaks a given system down into its simplest, most elementary components and shows how the organization of these simple parts produces the complicated behavior of the entire system. For example, in the case of a computer program, breaking down the system into its components, and breaking these components down into components of components, in the end leads to the most elementary components, to basic logical operations done on the symbols zero and one, which can be expressed physically as follows: zero is translated into a state of absence of electricity, and one is translated into a state of electricity. The interpretation of this translation from the symbolic system to the electrical system is that the physical system realizes the software or that the software is applied on the physical system (see also discussion in Rakover, 1990).

Cummins's functional analysis gave rise to several critiques (e.g., Griffiths, 1993; Millikan, 1989), which I shall not deal with here. My principal interest is to point out the two following problems, which touch on psychology. The first, which I shall call "complicating decomposition," gives rise to the following reservation. Cummins (1983) suggests that the decomposition of the complex system in the end extends to the simplest elements; but I suggest, based on observing the decomposition process as it in fact proceeds in scientific research, that in many cases the decomposition reveals elements that are not simple but actually extremely complicated, that is, components whose explanation is connected to a very complicated theoretical development. Cummins (1983) writes:

> Functional analysis consists of analyzing a disposition into a number of less problematic dispositions such that programmed manifestations of these dispositions amount to a manifestation of the analyzed disposition. By "programmed" here, I simply mean organized in a way that could be specified in a program or flow-chart. (p. 28)

In fact, the strategy of decomposition (from complicated to simple) was also proposed by Dennett (1979) as an explanatory diagram of anchoring the cognitive process to neurophysiology: the decomposition starts with a ramified system, passes to less complicated components, and finally comes down to the simplest elements, so simple that they can be replaced by a physical component with two possibilities: yes (there is an electric current) and no (there is not an electric current). To illustrate this research strategy, Dennett uses the metaphor of a homunculus, which describes the functioning of a mental mechanism: the decomposition advances from a very clever little man, who is broken down into smaller and less clever people, until the process comes down to tiny and stupid people, who without any difficulty can be replaced by a mechanical mechanism (i.e., here is a process that explains how to replace a mental mechanism with a mechanical mechanism).

In my view, the best example in which the decomposition strategy (from the complicated to the simple) works well is the computer: high computer languages can be wholly decomposed into languages ultimately approaching chains consisting of logical units (*and*, *or*, *not*), which are translated into two physical states: there is/there is not an electric current. But it is not so in other cases. We may look at the simple system of the flashlight, which can be broken down into the components bulb, electric wires, switch, battery. Now continuing the work of decomposition of these components, we shall very soon come upon not simple components but actually components that call for the most complicated explanations. For example, let us begin with the question of distribution of light, that is, the relation between resistance (the filament in the bulb) and the strength of the electrical current that passes through the filament (the bulb itself is empty of air), and from there we reach the question of the functioning of the electric current, composed of electrons whose explanation requires the use of a very unsimple theory, namely, quantum theory. Something similar may happen when we dismantle certain areas in the brain down to the extremely complex activity of the structure and functioning of a nerve cell.

The second problem, which I shall call "incomplete decomposition," yields the following reservation: in the computer, the passage from language to language up to the entire translation into logical units expressed in components of there is/there is not electricity is a complete translation (without loss or addition of information): but in mental and cognitive systems, the decomposition is not complete, and at times causes changes in the information, so that the components no longer represent the information that the system as a whole treated. To illustrate this notion, we shall look at the following sentences uttered by Ruth: (I) David *swept* all the dirt off the floor; (II) David *swept* all the soldiers from the enemy position. In both sentences the verb *swept* appears, but its meaning is entirely different in the two different contexts: in sentence (I) the context is housework; in sentence (II) the context is warfare. And a further example: The computer will always give the same answer to the question 2 + 2 = ? (i.e., the question will be translated perfectly from the human language into the computer language, consisting of logical units expressed by two electric states). But the meaning of this symbolic operation will be given by a human being, for example, if at issue here are oranges or perhaps containers each holding goods worth a million dollars.

For these reasons, among others, I don't believe that it is possible to sustain the proposition that the brain functions like a computer. It is hard to see how it is possible to decompose a certain cognitive system, for example, a system that handles recognition and identification of faces, into components, and components of components, which become increasingly simple down to a level at which it will be possible to exchange these wholly nonsophisticated components for simple neurophysical units, such that the mind will be comprehensible by means of the brain's neurophysiological activity. In my view, in the brain there seems to be "something" extremely important (that changes us from nondescript robots/zombies, to living beings/humans), some kind of system that supplies consciousness to some of the actions that the brain performs, namely, to conscious processing of part of the huge collection of stimuli absorbed in the brain. I have no idea if this speculation is correct, and in short, I repeat, a mind/body theory has not yet been discovered.

THE NEW-MECHANISTIC MODEL OF EXPLANATION

For the last twenty-five years, a group of researchers has been developing a relatively new approach to giving explanations, based on the conception that began to bloom already at the time of the French philosopher Descartes. It holds that the world is to be understood mechanistically (see Craver & Tabery, 2016). This approach, which Craver and Tabery call the "new-mechanism," is found earlier in studies by modern philosophers such as Cummins (1975, 1983) and Salmon (1984, 1990). As distinct from previous conceptions of explanation (particularly those reviewed in chapter 2), which centered on physics, this approach was inspired by the special sciences, such as biology, neurophysiology, and psychology. (In fact, it is gradually becoming clear that new-mechanism has spread its wings over many and varied research fields such as chemistry astrophysics, genetics, evolution, the cognition sciences, social sciences, and computer sciences. This approach appears to be steadily capturing many hearts and is becoming an accepted explanatory framework; I shall call it the "new-mechanistic explanation," and it is of great importance for understanding explanations proposed by cognitive psychology [see Bechtel, 2008a, 2008b, 2008c].) What then is this new-mechanism?

Bechtel and Richardson (1993) propose explaining complex systems by means of the new-mechanistic explanation, which they define thus:

> By calling the explanations *mechanistic,* we are highlighting the fact that they treat the systems as producing a certain behavior in a manner analogous to that of machines developed through human technology. A machine is a composite of interrelated parts, each performing its own functions, that are combined in such a way that each contributes to producing a behavior of the system. A mechanistic explanation identifies these parts and their organization, showing how the behavior of the machine is a consequence of the parts and their organization. (p. 17)

Bechtel and Wright (2009) apply the new-mechanistic approach to psychology as follows:

> A mechanism is simply a composite system organized in such a way that the coordinated operations of the component parts constitute the mechanistic activity identified with the explanandum. . . . The mechanism as a whole is comprised of component parts, the orchestrated operation of which constitutes its function(s). Not infrequently, the parts of a mechanism are themselves mechanisms consisting of component parts at lower levels, which implies that mechanisms are intrinsically hierarchical. (p. 119)

(See similar approaches to the new-mechanistic explanations in Craver, 2001; Machamer, Darden, & Craver, 2000.)

As a simple example, I shall apply the new-mechanistic explanation to the operation of the flashlight (it is much simpler than the weather, or animal or human behavior). To answer the question, "How does this instrument work?" one performs the following theoretical–empirical analysis.

1. *Decomposition into parts*: One decomposes the instrument—the flashlight—into the following parts: switch, battery, electric lead, bulb.
2. *Explanation of every part*: The explanation of every part is based principally on the laws of electricity. When the switch is operated, the electric circuit is closed. The current passes through the filament in the bulb (e.g., a tungsten wire), heats it, and as a result the bulb emits light. And now we approach the explanation of the action of each and every part: the action of the battery on the basis of chemical and physical processes, the action of the filament as electrical resistance, and finally the laws of electricity concerned with connection in series as distinct from connection in parallel. All these explanations are based on known physical theories (electricity and chemistry).

3. *Explanation of the interaction between the parts*: We explain the connection between the different parts as a process of conversion of energy: chemical energy, which converts into electrical energy, which converts into heat, which emits light. This energy conversion can be calculated precisely and explained by means of the appropriate laws and theories.

I would like to stress here three important points regarding the explanation of the flashlight's operation:

1. *Explanatory independence*: We break down the overall system into its components, its parts, and explain each part separately—for example, how the battery or the bulb works in a flashlight, or to take another example, how the engine, the electrical system, or the cooling system works in a car. We assume that each and every part is independent explanatorily. That is, we assume that despite the important interaction between the parts, an interaction that determines the action of the system as a whole, each and every part, each and every component, can be given an appropriate and adequate explanation. Furthermore, the flashlight and its components are independent of the surroundings since they do not affect the flashlight's operation except in extreme cases (the instrument is broken or burned).
2. *Explanatory relevance*: Consider the motion of a car: each and every part in the overall system renders a number of products; for example, the motor causes the mechanical motion that rotates the car's wheels, but it also generates noise and noxious gases. For the explanation of how the car works as a single entire unit, the noise and the gases are not only irrelevant, but are also a kind of negative products whose adverse effect must be reduced to a minimum. What is important for the action of the car is the relevant fact that the engine converts energy associated with burning benzene into mechanical energy, which can be harnessed for the purpose of setting the car in motion. That is, although every part is amenable to an explanation independently of another part that contributes to the overall system, each part has an important property that is relevant to the proper action of the system as a whole.
3. *A uniform kind of explanations*: The explanation described above is based on diverse physical theories that are applicable in the specific case of the flashlight too. The unifying property of these theories is that they are all mechanistic explanations. This property allows us to attain a uniform explanation of the entire system. This goal is realized by the rules of energy transformation, by means of which we pass from one computation of energy to the next. That is, these rules of transformation allow us to perform a precise calculation of the results that are measurable empirically, so that we can know exactly the amount of current that will pass

through the filament in the bulb and as a result know the amount of heat and light created in the bulb. That is, the use of these theories allows us to calculate precisely the amount of energy of each and every component, and the order of the conversions of battery-produced energy into energy produced by the flashlight bulb.

As may be seen from the foregoing discussion, and from the illustration of the flashlight, four characterizations arise out of the concept of the new-mechanism:

1. *The phenomenon under study.* Usually the phenomenon is a complex system of activity—for example, the light shed by the flashlight, or as a much more complicated example from research in psychology, the mechanism of face perception and recognition. In a substantial number of cases, the mechanism (which is perceived as causal chains) creates an end-state, such as light in the case of the flashlight, or an end-response (recognizes or does not recognize the face) in the case of the conceptual system: David, who identified his friend Johnny, said, "Hey Johnny, it's been a long time since I saw you last. Wow—you haven't changed a bit." In such cases, the mechanism offers a (causal) explanation by whose means one understands how the given behavior was produced. In another set of cases the mechanism constitutes an elementary component of the studied phenomenon itself, for example, courtship between males and females (not only in human beings), a learning process, a process of perception, recognition, or recall. In these cases, the studied behavior is understood by decomposition of the phenomenon into its parts, including their functions and their organization.
2. *Components of the mechanism.* For example, in the case of the flashlight, the components of the system are switch, battery, electric wires, bulb; in the case of the mechanism of the human memory, the components are sensory register, short-term store, and long-term store (see Bechtel & Wright, 2009).
3. *Functions.* Each component of the mechanism has a characteristic function. In the case of the flashlight, the switch turns the electric circuit on/off; the battery supplies the electric current; the electric wires carry the electric current (isolated from its surroundings); the bulb supplies heat and light as a result of the heating of the filament (incandescent wire) inside. In the case of the memory system, the sensory register (e.g., the visual store) retains all the information in the stimulus for fractions of a second; the short-term store retains information that has undergone transformation into sounds (language) for tens of seconds; and the long-term store retains information for decades. In addition, inside the mechanism of the human

memory, more processes operate, such as coding, storage, and retrieval, and processes associated with decision-making that control these memory stores.

4. *Organization.* The components of the mechanism and their functions include processes of control, organized in the mechanism in such a way that the studied behavior will be produced (light in the case of the flashlight, and recognition of the stimulus, such as the face, in the case of human memory). In the example of the flashlight, this mechanism's components are organized not in parallel but in series, so that turning on the switch will pass the electricity through the filament in the bulb and will create the required amount of light. In the case of human memory, this mechanism is arranged in such a form that will meet several theoretical requirements—for example, the mechanism will respond to an external stimulus of a certain kind (and not to an internal, imaginary stimulus), but in essence this mechanism is so constructed as to be able to explain successfully several empirical findings, certain important behavioral phenomena. An example is the empirical finding that an individual can remember all the visual information if he is examined by a special technique in a range of hundredths of a second after presentation of the visual stimulus (retrieval is from the visual store); the empirical finding that an individual is able to remember a small number of details in a range of tens of seconds immediately after their presentation (retrieval is from the short-term store); and the empirical finding that learning new material impairs retrieval of old material from the long-term store. Clearly, if the organization of memory were different—for example, if the order of transfer of the information were not from the sensory register to the short-term store and thence to the long-term store—it would not be possible to explain some of the above findings. For example, if the sensory register received the information only after it had entered the long-term store, it would not be possible to explain the stunning finding that all visual information is stored in the sensory register for a fraction of a second only.

In view of the foregoing, it seems, then, that the flashlight mechanism can be treated in two ways: on the one hand, this mechanism serves as a specific explanation for the empirical observation about producing light by a flashlight; on the other hand, this mechanism itself may be viewed as a phenomenon to be explained by appeal to physical theories (electricity and chemistry) that are well-founded and well-known. But not only that; ultimately, these physical theories have guided researchers and inventors to construct and develop the flashlight (in 1899 David Misell registered a patent for a flashlight).

Application of the New-Mechanism Model of Explanation to Psychology

Given the flashlight analysis, can a similar analysis be conducted on human behavior such as perception and memory of faces? My answer is negative. This response is based essentially on a theoretical/empirical fact (which I have already underlined): still not found is a theory that explains consciousness, awareness by means of the brain's neurophysiological processes, namely, a theory of consciousness/brain's neurophysiology or a mind/body theory (see Rakover, 2007, 2011/2012a, 2011/2012b; Uttal, 2014).

Therefore, I do not believe that the following proposition by Piccinini and Craver (2011) describes the proper relation between psychology and neurophysiology:

> Thus, if psychological explanation is functional, as so many people assume, and psychological explanation is worthy of its name, then psychological explanation is mechanistic. Once the structural aspects that are missing from a functional analysis are filled in, functional analysis turns into more complete mechanistic explanation. By this process, functional analyses can be seamlessly integrated with mechanistic explanations, and psychology can be seamlessly integrated with neuroscience. (p. 308)

The new-mechanistic explanation for a complex system in the natural sciences is founded on causal processes (e.g., conversion of energy) that underpin the interactions between the components of the system (e.g., Bechtel & Richardson, 1993; Craver, 2001; Machamer, Darden, & Craver, 2000). It is very difficult to apply this approach in psychology. As mentioned above, I don't think that today there is any philosophical approach or scientific theory that is generally accepted and that does not spark sharp criticism that is able, in the framework of the scientific game rules, to bridge the mental and the physiological, even though this connection of mind/body is a common and characteristic phenomenon in animals.

In the new-mechanistic explanation, the phenomenon being investigated is analyzed by means of mathematical, probabilistic, physical, chemical knowledge. By contrast, daily human behavior is explained by means of practical (private, subjective) considerations. As an illustration, consider the following example concerning road rules. If David behaves according to the traffic regulations, we may say that David is behaving rationally. But let us change the situation a little: David is driving fast at three in the morning down empty streets because he is in a great hurry to get to the hospital, where, on account of a sudden and massive heart attack, his mother has been admitted. Now, shall we not justify David's speeding above the allowed limit? Is David behaving rationally? On the one hand, he is breaking the law, endangering his life, thus reducing his chances of seeing his mother recovered; on the other hand, the chances of an accident at this early hour are very small, while the chances of his reaching the hospital after his mother has died are, sadly, high. The decision to drive fast or slow is therefore subject to David's considerations, beliefs, and feelings. That is, what is relevant depends on David's viewpoint. These mental states (will, belief, feelings) are hard to conceptualize neurophysiologically, so it is almost impossible to account for David's behavior by appeal to models of explanation based on the natural sciences' framework. For instance, it is hard to materialize in psychology the property of explanatory independence as it is realized in the case of the flashlight; it is not clear how the conversions of energy between diverse behavioral and mental components may be calculated—namely, it is not clear how the interactions between these components may be described.

In this respect, compared with the explanation given in the natural sciences, the explanation for human behavior is partial and not complete. In the case of the flashlight, the explanation constitutes a single whole (because it details the conversions of the energy, that is, the mechanism that provides the light when the switch is activated and closes the electrical circuit). By contrast, in the case of David's waving as a gesture of taking leave of Ruth, we are unable to detail wholly the mechanism responsible for his behavior from the moment David catches sight of Ruth's face until the moment he performs the waving behavior.

Despite these differences between the natural sciences and psychology, it is worth noting that in the natural sciences too there are complex phenomena that are difficult to comprehend by application of the "decomposability" strategy: breaking down the parts of the system, and identifying the functioning and the organization of the components. In fact, phenomena may be aligned on a scale from simple (such as the flashlight) to the most complex (such as the human brain). Phenomena of the first kind are decomposable, and the functioning of their components is readily identifiable. But phenomena of the latter kind are typified by a large multiplicity of interactions among the components, a situation liable greatly to hinder the application of the decomposability strategy and understanding of their action. Nevertheless, compared with research in the natural sciences, the work of the scientist attempting to understand the behavior of animals (especially humans), involving the mind/body interaction, is much harder.

Bechtel and Richardson (1993) suggest two complementary strategies for an explanatory decomposition of a complex system: the analytic strategy, whereby a physical component in a system is isolated, and its functioning is examined, and the synthetic strategy, whereby the researcher proposes a scheme of behavior consisting of different parts as a hypothesis for the functioning of the system, a hypothesis that is tested empirically. While the analytic strategy takes a bottom–up approach (i.e., from an attempt to understand a system on an explanation track that goes from the basic components to the structure of the whole system), the synthetic strategy takes a top–down approach (i.e., from an attempt to understand a system on an explanation track that goes from the structure of the whole system to its basic components).

A number of researchers have attempted to apply the new-mechanistic model of explanation to psychology and to the cognition sciences (e.g., Bechtel, 2008a, 2008b, 2008c, 2009; Bechtel & Wright, 2009; Piccinini & Craver, 2011; Wright & Bechtel, 2007). Here, I do not intend to review all the interesting ideas proposed in these studies, but to pause at a few points that touch on my approach as clarified above, where I reasoned why I do not accept the proposal of Piccinini and Craver (2011): although the development of an explanation by means of a description of a cognitive/mentalistic mechanism may contribute to understanding several cognitive/behavioral phenomena, I do not see how this important research tool (which, as stated, was developed essentially for research in biology and neurophysiology) is likely to explain behavioral phenomena interwoven and imbued in consciousness.

In his book, Bechtel (2008c) reviews briefly the development of the new approach to mechanism. He notes that this approach, which is subject to characterization as naturalist, pays great attention to researchers' actual practice, and in that way runs counter to the previous approach, which developed explanatory models on the basis of thought/logical analysis. As an example of this trend, one may present research in psychology. On the basis of the appropriate literature, Bechtel and Wright (2009) comment that psychological research connected to empirical psychological laws declined dramatically from 1990 to 1999, and in its place arose the use of explanation by means of a description of cognitive mechanisms. (In this regard one may say that while the earlier approach tended to construct normative models, the present approach tends to descriptivity.)

The physical mechanism as in the example of the flashlight is wholly explainable by appeal to the appropriate physical theories: the mechanism developed for biology and neurophysiology is not explainable that way, inter alia because a mechanism of this kind is greatly influenced by the environment in which it is immersed. This is a mechanism that operates in a living organism—namely, it is a living mechanism, wherein the interaction between it and the environment is vital for its survival. By means of the development of a living mechanism of this kind, Bechtel (2008c) continues, he reaches the mental mechanism. The basic difference between the new-mechanistic explanation developed out of research in the natural and life sciences, and that which deals with biological phenomena, and the mental mechanism is that the latter is a mechanism for information processing (IP) (see Figure 3.1).

As such, the mental mechanism must take into account that the components of the mechanism represent the world, that is, the representations of the particular contents of the world in which the individual exists, and also the operations performed on these representations.

The IP mechanism is constructed of certain components that join up in relations of information flow, which undergoes various processes according to certain processing rules. (It is important to stress that each component of the mechanism can be decomposed into subcomponents.) Usually a mechanism of this kind is described by a diagram of boxes and arrows, where the boxes refer to different cognitive components, and each arrow describes the information flow from one component to another. The box or the cognitive unit is a component possessed of a certain capability of information processing (i.e., a rule of processing, an operation done on the representations, the symbols). (Here, I do not intend to distinguish models constructed on a pattern of classic software from models constructed on the basis of neural networks.) While the researcher in biology can identify the operations of the mechanism's components by measurement of material transformations (e.g., electrical and chemical changes in the brain), researchers in information processing cannot: one cannot measure feelings, emotions, and thoughts that pass through the individual's head. This cognitive research may be somehow fostered by new techniques of brain imaging, which can measure, for example, accelerated action of passage of blood in certain brain areas, when the individual is busy performing cognitive functions. Yet here, too, the complexity is great, because the same brain area is able to handle several different functions and behavioral phenomena (see Bechtel, 2008a, 2008c).

As can be seen in Figure 3.1, the explanation of the given behavioral phenomenon is accomplished by the IP mechanism as follows: a certain stimulus state (i.e., input) operates on the cognitive mechanism that processes the information passed to it from the input; as a result, the mechanism produces/creates, a certain prediction (i.e., the mechanism produces output). When the prediction matches the given behavioral phenomenon, it is postulated that

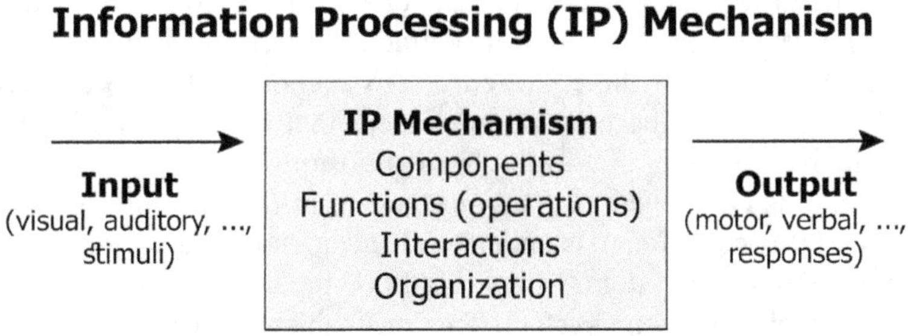

Figure 3.1 The Information Processing (IP) Mechanism.

the given mechanism has succeeded in its task: a satisfactory explanation has been given for the behavior under study. When there is no match between what the mechanism produced and reality, the explanation has failed, and the mechanism has to be changed or a new mechanism has to be sought/constructed.

Bechtel (2008c) sums up the difference between a mechanistic explanation and a mentalistic explanation thus:

> Mechanistic explanation consists in representing (verbally or in diagrams) the parts, operations, and organization, and showing (often via mental or computational simulations) how such a mechanism realizes the phenomenon in question. Mental mechanisms share these features, but operate not just to transform physical substrates but to process information. Hence, we will find that mental mechanisms are often most usefully described in terms of the content they carry, not their intrinsic physical features. (p. 48)

In light of this application of the new-mechanistic model of explanation to psychology, I would like to conclude the present review with the following three points: the first focuses on my experiment (Rakover, 2013) to develop an explanatory model of perception of an inverted face by the development of a cognitive mechanism, a development that will bring me directly to the second point. This will focus on the relation of the present new-mechanistic approach to consciousness. The last point will concentrate briefly on whether the new-mechanistic approach is likely to replace the traditional approach of explanation by means of laws (or well-founded theories).

First, the model I published in 2013, the face-scheme incompatibility (FSI) model, sets out to contend with a large collection of empirical findings on a phenomenon called the Face-Inversion Effect (FIE). This phenomenon is constructed of two kinds of empirical findings: according to the first kind, named in that article the "face(UI) effect," recognition of a face presented inverted (hair below, chin above) is much lower than recognition of a face presented upright (hair above, chin below); according to the second kind, called in the article the "face/object(UI) effect," inversion impairs face recognition far more than inversion of nonfacial objects (e.g., buildings, trees, cars). The basic idea of the FSI model is to anchor the FIE to the development of a scheme for the perception of faces that have appeared before the individual since birth in an upright, normal posture.

The model is conceived as a submechanism in the general visual mechanism, just as this mechanism is understood and anchored to the general framework of the cognitive system, which is built on the metaphor of the computer, the information-processing system. This mechanism is constructed of different components that process visual information. Rakover (2013) writes:

> The components are organized and interact so as to produce the phenomenon under investigation. It is proposed that the description of the mechanism (the components and their activity) mirrors the actual activity that occurs in one's cognitive system when one processes facial information. While the mechanism's activity is automatic and beyond one's awareness, the phenomenon to be accounted for—the face(UI) effect—involves one's consciousness. (p. 674)

The model is constructed of three stages: in the first and second stages it is explained how in the cognitive system a scheme of an upright face develops from visual-facial stimuli; in the third stage it is explained how this scheme handles a new facial stimulus, and it is assumed that the facial scheme is that which bestows on the facial stimulus face-form meaning that finds expression in the conscious experience of perception of a human face. To the question of how the scheme imparts to a visual-facial stimulus a conscious meaning of face perception, Rakover's (2013) model gives no answer. The reason is that research on the subject has still not succeeded in discovering a theory that will explain the connection between the neurophysiology of the brain and the marvelous phenomenon of consciousness and awareness (as argued above, the mind/body theory has not yet been discovered). And this is the second important point that I wish to emphasize.

Secondly, in physics and in the natural and life sciences the phenomenon under study has an explanation given by the causal mechanism that *produces/creates* the given phenomenon, where the terms *produce* and *create* have a precise causal meaning, for example, transformations of energy, chemical changes, and the like; but in psychology the terms *produce* and *create* have no such clear meaning, because the mind/body theory has not yet been discovered. Nor does the fundamental concept in cognitive psychology, the concept of information, have any clear meaning, and in fact the notion serves as a catch-all for a medley of phenomena that may be conceptualized even by association as information, from thoughts, images, and memories of words and pictures, to motor behaviors (see Palmer & Kimchi, 1986).

In chapter 7, which closes his book, Bechtel (2008c) refers briefly to the question of consciousness. He notes that the philosophers of mind give the name *qualia* to the qualitative characterization of the individual's personal experience, and he states, without offering a solution to the question of consciousness, that whether an explanation is found for the concept of qualia or not, this concept will refer exclusively to the individual's sensory experience and only the individual herself will be capable of evaluating it.

Thirdly, in his 2010 article, Leuridan examines the question whether the new approach of mechanism can take the place of a law of nature (a strict law, a law without exceptions) in scientific explanation. His answer is negative. He notes, in fact, that laws of the strict kind are hard to come by in the natural, life, and social sciences. What is available are pragmatic laws, namely, empirical generalizations or stable regularities. Regarding these generalizations, Leuridan's analysis shows that not only is mechanism not about to replace them, but also mechanism itself depends on them. It arises that the mechanism notion is based on a certain regularity, an empirical generalization, from the very fact that mechanism expresses organization and interactions among its components so that the studied phenomenon will be explained as produced by the mechanism under discussion. In other words, the mechanism is organized, is arranged, and reveals in its foundation a regularity whereby in certain conditions it produces the phenomenon under study (this criticism, as may be expected, did not remain without a response: see Kaiser & Craver, 2013).

To conclude this chapter, it is appropriate here to anticipate and say that compared with Bechtel's (2008a, 2008b, 2008c) approach, in the following chapters I shall suggest a different interpretation of the concept of mental mechanism. Bechtel proposes that a mental mechanism is one that processes information—essentially on the nonconscious cognitive level; I maintain that the action of the mental mechanism is located on the conscious level—namely, the interactions among this mechanism's components take place consciously. I shall illustrate this with the teleological explanation or the will/belief explanation. For example, David traveled to Tel Aviv; how do we explain this journey? The explanation is as follows: David *wished* to hear Mozart's opera *Don Giovanni* and *believed* that a journey in his car from his place of residence, in Haifa, to the Tel Aviv Opera House at the appropriate time would realize his will, so he drove in his car to Tel Aviv. In this case then, the explanation is based on a mental mechanism in the framework of the rational considerations of the individual, who examines in his consciousness the practical means of realizing his will: for example, David checks the cost of implementing these means and wants to choose the means whose price will be minimal. He asks himself if the cost in time and gasoline of the journey to Tel Aviv will outdo the enormous pleasure of seeing this wonderful opera: this cost would probably be the decisive factor if *Don Giovanni* were being staged at the Metropolitan Opera in New York. In light of this example, it is reasonable to suggest that the mental mechanism is composed of conscious mental components (will, belief) that play an important part in a conscious rational process (practical considerations) that in the end leads the individual to perform an action (behavior) that satisfies his will.

Many researchers have distinguished two kinds of cognitive systems: System I—automatic, nonconscious; and System II—deliberate, conscious (e.g., Kahneman, 2011; Reese, 1989; Evans & Stanovich, 2013; however, the results of Weidemann, Satkunarajah, & Lovibond, 2016, contradicted this dual-system approach). Reese (1989) writes: "Cognitivists distinguish between controlled or effortful cognitive processing and automatic cognitive processing. Effortful processing is the deliberate, consciously controlled use of rules; automatic processing is not deliberate and not consciously controlled" (p. 6).

It is reasonable to suppose that the individual is not aware of some of the processes operating in the cognitive system, while he is conscious of some of them. Some researchers maintain that the individual is aware only of the final outcome of the cognitive process, but this matter is still controversial (see Reese, 1989). David, for example, is fully conscious of the considerations for and against the drive to Tel Aviv, but is not aware of the associative process that caused him to suddenly remember the possibility that the journey to Tel Aviv was liable to create tension between himself and his boss, who had asked him to stay behind at the office on the very evening of the performance on account of the work overload. And a further example: I can be aware that 2^2 equals 2×2 equals $2 + 2$, but I am not aware of the processes concerned with retrieval from memory of these arithmetical symbols and their meaning—these are operations wholly hidden from my inner eye. Considering this, I suggest that the mental mechanism has different degrees of consciousness, control, and weighing up: from one extreme, *the automatic and not-conscious*, where the result of the not-conscious action of this mechanism appears suddenly in the conscious, to the other extreme, *the conscious and the deliberate*, in which the individual is aware of some of the rules by means of which the considered action, along with its outcomes, is carried out. I propose calling the first mechanism the cognitive mechanism, and the second the mental mechanism, because in cognitive psychology the former process is not necessarily connected to conscious processes, while the mental concept is connected phenomenologically to the consciousness concept.

DISCUSSION: DO WE HAVE UNIQUE EXPLANATION MODELS BEFORE US?

The present chapter has discussed three kinds of models of explanation that are of great use in psychology and the social sciences: rule-explanations, teleological explanations, and new-mechanism explanations. The question up for discussion here, and which I have also treated above, is this: Are these three models unique—namely, does each of them stand alone without it being possible to reduce them to any other explanation model? Following these discussions, my answer to the question of "specificity of the explanation" was affirmative: these models stand on their own feet, and it does not appear that other models of explanation are able to embrace them.

Hempel proposed the D-N model as a general model that spread its wings over different explanations in many areas of scientific research, for example, causal explanations and explanations in historical research. The foregoing discussion showed that the Hempelian model encountered many difficulties in treating causal and historical phenomena, and I believe that I shall not be far wrong if I say that most researchers do not see the D-N model as covering all the areas that Hempel intended.

In consequence of the above discussion, the following important question arises: What does it mean that explanatory model A is not unique because it is reducible to explanatory model B? In other words, what does it mean that explanatory model A is reduced to explanatory model B? Can it be posited that reducing explanatory model A to explanatory model B is parallel to reducing theory X to theory Y? I suggest a negative answer because an explanatory model does not have a structure parallel to that of the theory. One of the differences between an explanatory model and a scientific theory is this: while an explanatory model may be viewed as normative—intended as a guide to what is to be done in order to propose a scientific explanation—a scientific theory is viewed as descriptive—a theory meant to describe processes in reality. (My approach to this subject is realist and not instrumentalist.) Theoretical reduction is achieved when it is possible to derive a description of the reduced theory, T_A, from a basic, reducing theory T_B, by means of bridging laws. For example, it is possible to show that Galileo's law and Kepler's law (the reduced theory, T_A) are reducible to Newtonian theory (the basic theory, T_B). Is a process of this kind widespread in the literature on explanatory models? To the best of my knowledge, the answer is negative: I have nowhere come across an attempt to infer a description of explanatory model A from explanatory model B by means of a few bridging laws. In fact, when an attempt was made to reduce one explanatory model to another, this theoretical attempt proved to bear the character of an exchange (or translation) of one scientific theory for another. I shall illustrate this by means of the theoretical attempt to translate a teleological explanation into a causal: the attempt of Hull (1952) and Spence (1956) to propose a behavioral translation (based on Stimulus-Response [S-R] concepts) into Tolman's (1959) mentalist/cognitive concept of expectancy.

A theoretical/empirical debate developed between Hull and Tolman stretching over many years: Tolman argued that a hungry rat learns the map of a maze at the end of which, in the goal box, it expects to find reinforcement (food); Hull maintained that all the rat learns is a chain of S-R that leads it from one end of the maze, the start box, to the other end, the goal box. Evidently, Tolman used a teleological explanation—the rat expected a reinforcement at the end of the course, and Hull used a causal, mechanical explanation—a chain of stimuli and responses that impelled/carried the rat from one end of the maze to the other. To apply the causal explanation to the rat's behavior in the maze, Hull and Spence translate the concept of expectancy (a hypothetical concept that is not objectively observable) into concepts of stimuli and responses (S-R, which are objectively observable). The underlying idea is to base this translation on two learning processes, well-founded empirically and theoretically: classical conditioning and stimulus generalization. And thus the expectancy concept is translated into S-R concepts: through classical conditioning the rat's response in the goal box (where the hungry animal got food) connects to the environmental stimuli of the goal box (the theoretical description is more complicated, but I shorten it here for simplicity's sake). The rat's response in the goal box is marked R_G and the stimuli in this box are marked s_g. The classical conditioning causes an associative connection: ($s_g - R_G$) in the goal box. And now, by the stimulus-generalization process stimuli similar in the maze to S_G, marked s_g, whose similarity to S_G constantly decreases with greater distance from the goal box, there arise in the rat goal responses similar to R_G, marked r_g, whose strength continuously decreases with greater distance from the goal box. These learning processes being assumed, ultimately a chain of stimuli and responses of the rat is obtained, connecting the start box (which we shall mark S_B) to the goal box. This chain has the following structure: $S_B - (r_g - s_g) - R_G$, where the symbols in square brackets indicate a chain of hidden stimuli and responses that lead the rat from the start box (S_B) through the maze to the goal box, S_G, where it obtains food and presents the R_G (see detailed description and discussion in Rakover, 1990). What does this theory (which we call the "($r_g - s_g$) theory") manage to achieve? I would like to highlight two things: first, Hull and Spence succeeded in explaining the rat's behavior in the maze without recourse to expectation, a subjective mentalist concept that is not publicly observable in people (not to mention animals such as a laboratory rat); secondly, a proposal is offered that translates/converts, the concept "expectancy" into the behavioral concepts ($r_g - s_g$) through a process firmly anchored to scientific methodology. In short, expectation is translated into a chain of stimuli and responses that lead the way from the moment of the appearance of the stimulus that arouses the expectation to the moment of its realization.

Does this theoretical proposition, elegant as it may be, reduce the teleological explanation model to the causal explanatory model? My answer is negative. If we perceive the explanatory model as a procedure, namely, as a set of directions for giving explanations in science, then ($r_g - s_g$) theory is clearly not such a set; therefore, it does not reduce the teleological explanatory model to the causal.

Here the following question arises: So what is the procedure whereby reduction between models (procedures) of explanation may be shown? My proposal is this: if the directions of explanatory model A are included in the directions of explanatory model B, then it will be possible to say that explanatory model A has been reduced to explanatory model B. As an example of this kind of procedural containment, we may look at the direction "Stop at the red light." This is a broad instruction that also includes all the following instances: at a red light, the pedestrian crossing is not to be crossed; at a red light, stop the vehicle (donkeys, horses, bicycles, motorbikes, cars) at the T-junction; at a red light, stop the vehicle at the intersection (+-shaped). By this approach, we may say that the $(r_g - s_g)$ theory has not illustrated a reduction of one explanatory model to another, but a theoretical competition—an alternative theoretical explanation, subject to an empirical test (and in fact this theory has received additional empirical confirmation; see Rakover, 1990).

Even if we regard the $(r_g - s_g)$ theory as an attempt to replace explanation by means of a future event (reinforcement at the end of the maze) with a causal explanation (a past event that causes an occurrence in the present or future)—that is, an explanation based on the use of memory of what was learned in the past as a cause explaining the hungry rat scurrying through the maze—this will not be some kind of reduction of one explanatory model to another, but substitution of a theory of one kind by a theory of another kind. In this sense I do not see what Salmon (1990) describes as constituting a procedure of reducing a teleological explanatory model to a causal explanatory model; rather, it is an exchange of theories:

> The basic idea of a consequence-etiology is as ingenious as it is simple. A particular bit of behavior B occurs because B has been *causally efficacious in the past* in achieving a goal G. A cat, hunting for prey, is clearly engaged in goal-directed behavior. It stalks in a typically catlike way because such stalking has resulted in the procurement of food. It is not caused by the future catching of this particular mouse, for (among other problems) in this instance he may not succeed in catching his prey. But such behavior has worked often enough to have conferred an evolutionary advantage on the members of the species. (p. 111)

What Salmon proposes in this passage (and in more passages that summarize literature on the subject) is relatively simple: instead of building a theory based on a future event that explains an occurrence in the present (a teleological explanation that modern science does not accept, among other things because we have no notion of what future event is likely to influence, and how it is likely to influence, the occurrence of a certain event in the present), a theory has to be constructed whereby past experience (stored in the memory, for example) will constitute the causal factor for behavior in the present. To my mind, this proposal is not a reduction of a teleological explanatory model to a causal explanatory model, but substitution of one theory by another. In this sense it is possible to see evolutionary theory, which explained the development of certain biological and behavioral qualities by a process of natural selection, as a theory that replaced a previous theory that was anchored to a religious approach, and which used future events to explain the occurrence of events in the present.

The teleological/religious theory was based on the following argument: because the world, nature, and reality behave in such an orderly and organized manner and not in a chance and chaotic way, a perception necessarily arises of an intelligent omnipotence that has planned everything in order to realize a certain goal. Similarly, we tend to answer the enthralling question, "Why was prehistoric Stonehenge erected in southern England?" with the statement that Stonehenge was planned to serve as a stellar observatory. Contrary to this tendency to suggest a teleological explanation anchored to the will of the omnipotent, Darwinist theory proposes an explanation that accords with the methodological requirements of science: Darwinist theory showed that very long-lasting processes of natural selection and the survival of the fittest to the changing environment are a good explanation for all developmental phenomena in our world, such as races, species, and interesting qualities in plants and animals (including humans).

Chapter 4

Reflections on Models of Explanation

Discussion of the fundamental notions of models of explanation reviewed in chapters 2 and 3 led me to contemplate two important questions. First, can these models be tested empirically? Second, from these models can certain properties, characteristics, and constraints be extracted that will constitute a theoretical infrastructure of a new general explanatory model for psychology? The present chapter will be the basis for the explanation models developed in chapters 6 and 7.

MODELS OF EXPLANATION AND EMPIRICAL TESTING

As the survey of the explanatory models indicated, some of them (especially those reviewed in chapter 2) underwent sharp criticism based on several interesting empirical illustrations. These showed that they were too broad because they contained explanations not perceived as scientific (the models were not a sufficient condition for a scientific explanation) or too narrow and did not contain explanations considered scientific (were not a necessary condition). For example, Salmon (1990, p. 50) presents in his critique of Hempel's models this entertaining example: anyone who takes pills against pregnancy will not get pregnant. For an entire year Mr. Johnny Jones took anti-pregnancy pills. Conclusion: Johnny Jones did not get pregnant. This explanation fits the D-N model (and the I-S model), but it is clearly built on an irrelevant component (anti-pregnancy pills being taken by a man).

This illustration and others, particularly those reviewed in chapter 2, have the same interesting structure: they are based on events/occurrences, that is, empirical observations (real or possible), that have an adequate and known explanation. The explanation for observations of this kind is: (a) it is well-known and obvious to the public at large, is general knowledge—for example, in the case of Mr. Jones, everyone knows that anti-pregnancy pills are meant for women, not men; (b) it is an explanation brought to the knowledge of the academic scientific community. (A collection of illustrations that may be sorted into category (a) or (b) appears in Salmon, 1990, pp. 46–50.) Philosophers have generally used observations of this kind (which I shall term "empirical distinctions") to criticize models of explanation and to point out their Achilles' heel. For example, the anti-pregnancy pills illustration shows that the model cannot determine what relevant information is (that is, the model is not a sufficient condition for a scientific explanation).

The concept of empirical distinctions differs from that of "intuitions" or of "imaginary cases," by means of which standard philosophy builds arguments that evaluate theories. In contrast to the latter concepts, the empirical-distinctions concept concerns observations anchored to reality and to a theoretically acceptable explanation. A new group of researchers, who support the approach called "experimental philosophy," criticize standard philosophy in that it is not based on the experimental methodology prevalent in the social sciences (psychology) but on the products of intuition and imagination (e.g., Knobe et al., 2012; Knobe & Nichols, 2008). Experimental philosophy studies classic philosophical subjects, such as moral judgments, free will, and consciousness, through the research methodology of the social sciences. However, to my mind the subject of explanatory models is not similar to the research handled by experimental philosophy. It is hard to evaluate how an explanatory model—for example, Hempel's D-N model—can be tested by the usual empirical testing method (the Hypothetico-Deductive [H-D] method)—a point I shall discuss forthwith.

The use of empirical distinctions to query the efficacy and applicability of a given explanatory model elicits the following question: Are models of explanation similar to hypotheses or theories that can be put to an empirical test? My negative answer, which runs counter to the approach of Weber, Van Bouwel, and De Vreese (2013), needs a broad foundation.

These researchers argue that the three philosophers of science Hempel, Kitcher, and Salmon, who proposed important and interesting models of explanation (reviewed in chapter 2), did not present an empirical test for their models, that is, for the descriptive aspect of these models; they write this about Hempel's model:

> A first problem is that Hempel does not provide arguments for his descriptive claim. He does not provide a database with records of relevant opinions of a large representative sample of scientists from all disciplines all over the world (i.e., their opinions on what kind of understandings is important). However, that is what he should have done if he wanted to build a convincing argument for his empirical claim. (Weber, Van Bouwel, and De Vreese, 2013, p. 29)

And regarding Kitcher's model, which leads to scientific understanding by means of the unification concept, they write: "He has not surveyed a large sample of scientists of various disciplines, in order to investigate what their views on understanding are" (p. 31). And similarly about Salmon's causal model: "Like Hempel and Kitcher, Salmon did not try to collect empirical evidence" (p. 33).

In short, Weber, Van Bouwel, and De Vreese (2013) maintain that these three philosophers have failed in that they have not adduced empirical support for the descriptive aspect of their models for giving explanations in science (I shall soon discuss the normative aspect of these models). They write:

> We have identified a major shortcoming which they have in common: they do not gather empirical evidence for their descriptive hypotheses (this could be done by systematically investigating the writings of a large representative sample of scientists) and do not argue for their normative claims either. (p. 37)

The approach of Weber, Van Bouwel, and De Vreese raises this question: How may a model of explanation—for example, the D-N model—be tested empirically? In my opinion, two possibilities present themselves here: (1) according to Weber, Van Bouwel, and De Vreese, it is possible to test empirically whether most scientists (e.g., psychologists) use this model to explain their research results; (2) it is possible to test empirically what this model determines has to be done and see whether the test's results endanger the epistemological status of the present model itself. I shall discuss the second possibility first (because the conclusion that arises from this discussion carries important implications for the first possibility).

Possibility 2: An Empirical Test of an Explanation Model

The D-N model holds that to explain a phenomenon or given observation, a prediction has to be derived deductively from the premises, which contain at least one theory, one natural law, and certain real conditions. After the prediction is derived and the research performed, the question arises whether the prediction emanating from this model is equivalent to the phenomenon under study or not. If the prediction is identical to the observation, we may say that the observation has been understood because it was expected to occur in accordance with what was established in the model's premises. If there is no equivalence between the prediction and the observation (namely, prediction ≠ observation [$p \neq o$]), it arises that the theory, the law, in the model's premises have not succeeded in explaining the given observation.

Since according to possibility 2 we are concerned to test empirically the D-N model itself, let us look at a case where the prediction differs from the description of the observation under study, $p \neq o$. How can this finding be explained? (Here, I don't plan to enter into a discussion of Duhem's problem, which has to do with refuting the research hypothesis together with auxiliary hypotheses: see discussion in Rakover, 2003.) There are two possibilities:

A. Accordingly, we shall suggest that the theory proposed in the explanatory model is incorrect, has failed in its handling the given observation.
B. Accordingly, we shall suggest that the present model is incorrect, is not sound, and therefore $p \neq o$ has been obtained.

I believe that possibility B is not acceptable for the following reason. The D-N model constitutes an explanatory storehouse for a huge collection of empirical observations. In numerous cases the proposed explanation is not a success ($p \neq o$). In light of possibility 2, and according to the empirical testing (the H-D method) known to everyone engaged in conducting experiments, we should have dismissed the present explanatory model ages ago! Since 1948 (the year of publication of the D-N model), thousands of theories have been found that have not been able to explain thousands of empirical observations. But no one maintains that this model is worthless because a great number of theories or laws have come up with bad explanations. The empirical result ($p = o$ or $p \neq o$), which by the D-N model states whether a successful explanation has been presented, implies nothing as to the degree of correctness (truth) of the model itself. (One may also say that the finding $p = o$ implies nothing for the status of the explanatory model itself.) In other words, the explanatory success/failure of the D-N model cannot assess the model itself, but only the theory, the law, inserted in this model. So how can this explanatory model be assessed? By rational reasoning or by reasoning about the model's match to different research domains (see later).

Possibility 1—Extent of Use of the Model

To reinforce my negative answer about the extent of use of the D-N model and its implications for the status of the model itself, we can study the following hypothetical situation: We take the D-N model and use it to construct a hypothesis—the "use hypothesis": Using the D-N model (U_{dn}) for proposing explanations. This hypothesis states, for example, that researchers in psychology use this model a great deal in explaining their research findings. The reasoning behind this hypothesis may be this: assuming that the D-N model is indeed efficient and good, it is reasonable to assume that psychologists will use it a lot. According to the accepted empirical testing method in psychology, the H-D method, we have to infer from the U_{dn} under certain conditions a prediction that is empirically testable (e.g., researchers in cognitive psychology will make much use of the D-N model). This prediction can be tested by our handing out an "Explanation" questionnaire that probes these psychologists' views on the use of models of explanation. Let's go on, and assume that these researchers indeed know the range of explanatory models (like those reviewed in chapters 2 and 3) to the same degree they know a range of statistical models to analyze their research data, and therefore they can fill out this "Explanation" questionnaire intelligently.

And now we can go further and assume that the percentage of users of the D-N model is 3.5 percent, and that the statistical analysis found that this percentage was no different from zero—that is, the research hypothesis U_{dn} was refuted. What does this finding say? As stated above, if U_{dn} means that "researchers in psychology make much use of this model of Hempel's to explain their findings," then this hypothesis is refuted and found incorrect. Does a conclusion arise from this about the correctness of the D-N model itself? In negating this, I wish to emphasize the following points. It can be sensed at once that the fact that most psychologists do not use the D-N model says nothing about the question of the correctness/incorrectness of the model itself (namely, whether this model is true or false).

First, it can be suggested that the amount of use of any instrument (in the present case, an important instrument for giving explanations) attests nothing about the measure of its correctness, nor even about the measure of its efficacy. People use work tools (physical or mental) by habit, convenience, resistance to change and innovation, and suchlike reasons.

Secondly, it appears that the finding that psychologists do not use the D-N model has no direct implication for the model itself, for its correctness, but, as stated above, it does have a direct implication for the U_{dn}, that is, for the use hypothesis in which this model features as an important component. According to possibility 2, *what is refuted is not the explanatory model itself but the hypothesis on the extent of use of this model.* Thus, it emerges that the D-N model is indifferent to the empirical testing. When the D-N model yielded a prediction subject to confirmation/refutation by means of an empirical observation, it turned out that the confirmation or refutation of this prediction implied nothing for the model itself, but only for the hypothesis U_{dn} set in the model and by which the explanation is proposed.

Thirdly, how may we explain the surprise finding that only 3.5 percent of psychologists use the D-N model? To do so, we must use a model of explanation. It is reasonable to suppose that the D-N model itself cannot be used to explain this finding, because whoever believes that this finding has a negative implication for the model's epistemological status will not use it as a vehicle for giving explanations. So let us use another explanatory model (M_x). But a problem arises here: Who can ensure that M_x is indeed efficient and appropriate without conducting empirical research? Clearly, the likelihood is very high that at least in one test case the result will be negative (i.e., $p \pm o$). This means that M_x can't serve to explain the 3.5 percent finding, and likewise the other possible explanation models!

Fourthly, models of explanation seem similar to rules of behavior, conventional customs, whose purpose is to propose justified rules of action for certain areas of behavior. In the present case, the use of explanatory models is justified in that these models are anchored to rational thinking and to the methodology of science and hence have proven suitable for use as explanation schemes that provide answers to *why/how* questions in certain research fields. In essence, the D-N model is no more than a procedure that tells the researcher how to explain a certain phenomenon—and this alone is what the model is able to do. Being such a procedure, it cannot be criticized, assessed in right/wrong terms, only in terms of rationality and efficacy in adjusting the model to particular purposes. For example, use of rational, logical arguments, whose purpose is to show that the model creates inner contradictions or meaningless conclusions; use of arguments that show that the D-N model cannot use hypotheses, theories, or laws in certain research fields; or use of empirical distinctions, which show that the model is incomplete, is too broad or too narrow to manage the explanations that appear in science.

Explanation Models and Normativity

Researchers in decision-making distinguish three kinds of models for making decisions: normative, descriptive, and prescriptive (e.g., Bell, Raiffa, and Tversky, 1988; Keller, 1989). If we apply this triple classification to the field of explanatory models, we shall obtain in the present case this possible division: a *normative* model for giving explanations tells the researcher what he or she must do, what actions must be taken when he or she is about to offer an explanation for the given phenomenon, because these actions are rational and justified; a *descriptive* model tells the researcher what kind of explanatory model is used by researchers engaged in the research field of interest to him or her; a *prescriptive* model is meant to persuade the researcher with regard to improving his or her approach to giving explanations so that these will meet the rational and justified requirement of the normative model. Likewise, the assessment of these models varies according to their nature: Bell, Raiffa, and Tversky (1988, pp. 17–18) proposed that the *normative* model be assessed by the *theoretical adequacy* criterion, namely, by the extent to which the model is based on good and accepted rational justifications; that the *descriptive* model be assessed by its *empirical validity*, namely, by the extent the model suits the way explanations are given in science; and that the *prescriptive* model be assessed by its *pragmatic* value, namely, the extent it improves the rational considerations in giving explanations (namely, how closely it approximates a suitable and good normative model).

Still, these three researchers point out that distinguishing these three kinds of models is not easy, because at times their overlap is too wide. They also indicate (on p. 10) that a normative model is built when the model's creator "has an eye on the real world for inspiration." I interpret this phrase thus: among other things, the creator of the new explanatory model scrutinizes a scientist's actions when he or she proposes explanations or searches for a general rational interpretation that is likely to constitute a model/scheme for the generality of explanations proposed by scientists. I stress the phrase "among other things" because the proposal of the new model is influenced by previous proposals, by their critiques, and especially by the empirical distinctions presented against the previous models. For example, in respect of previous empirical distinctions, the M-I model contends successfully with the barometer and the flagpole and shadow illustrations by using counterfactual situations: in the first instance, changing the barometer reading, which in no way affects the onset of the storm; and in the second instance, asking if change in the shadow's length is possible without changing the height of the flagpole.

The three researchers, Bell, Raiffa, and Tversky (1988), illustrated their triple classification by means of the concept of transitivity. For example, normativity finds expression in the following proposition: if A is chosen relative to B, and B is chosen relative to C, then it is advised to accept that A will be chosen relative to C. This is an important assumption for the normative model in decision-making, and usually it holds when tested empirically as well—that is, a descriptive decision-making model will in most cases not err if it assumes transitivity. (Here, I do not discuss cases that do not uphold transitivity empirically; however, as a hint: think, for example, if the relation "win" upholds the transitivity rule in the national soccer league among competing clubs.)

In light of the triple classification (descriptive, normative, prescriptive), the question arises how to treat the models reviewed above. For a representative illustration, we can weigh up the D-N model once again. From the reasoning leading to the conclusion that neither the empirical test of the D-N model nor the measure of use of this model carries any implication for the correctness of the model itself, a reasonable possibility arises that this model has to be characterized as normative/prescriptive. Nevertheless, it is reasonable to assume, in keeping with the proposal of Bell, Raiffa, and Tversky (1988), that Hempel and Oppenheim created the present explanatory model with an eye on reality for inspiration, so it would be awkward to neglect the model's descriptive aspect. However, the critiques leveled at the model were not anchored to the empirical test of the use of this model, but were based essentially on empirical distinctions, whereby it emerged that the model is not a kind of a necessary or sufficient condition for a scientific explanation.

So now we ask: What does it mean that the D-N model is normative/ prescriptive? To my mind it means that this model tells researchers something like this: if you want to explain your research results, there are rational reasons and good justifications that make it worth your while to use the D-N model: find a well-founded relevant theory, *T*, apply it to your experimental condition, *C*, draw from these two (*T* and *C*) a clear prediction, and check how closely it matches the research findings. If you find a match, you will have succeeded in explaining your research observation by means of *T*; if you don't, then *T* is apparently not suitable, and you must amend it or look for another theory.

Explanation Models and Experimental Philosophy

As argued above, the present discussion touches on the debate about the new approach to philosophy, experimental philosophy (see Knobe & Nichols, 2008). But before discussing this subject, a preamble is in place: the above conclusion regarding the absence of epistemological implications of the empirical test for the D-N model should not be used to undermine the methodological status of the experimental approach to philosophy. Rather, the reverse: I believe that in quite a few areas (e.g., moral responsibility, moral norms, religious views, acquired knowledge), empirical research is likely to make an important contribution, connected, for example, to psychological processes that underlie these philosophical questions.

One of the most interesting arguments against experimental philosophy says:

> In a typical experimental philosophy paper, the evidence being gathered is about the percentages of people who hold various sorts of intuitions, but the theories under discussion are not about people's intuitions but about substantive philosophical questions in epistemology, metaphysics, or ethics. . . . How on earth could information about the statistical distribution of intuitions ever give us reason to accept or reject a particular philosophical view? (Knobe & Nichols, 2008, p. 6)

In reply to this question, Knobe and Nichols suggest that empirical findings should be used indirectly, as a comparative frame for a discussion of the given subject. They posit, for example, that the most important goal of experimental philosophy is "to determine what leads us to have the intuitions we do about free will, moral responsibility, the afterlife. The ultimate hope is that we can use this information to help determine whether the psychological sources of the beliefs undercut the warrant for the beliefs" (p. 7).

It is reasonable to assume that a philosophical debate—on moral responsibility, for example—will gain an important contribution from the findings of experimental research on the subject, because ultimately moral responsibility is created and constructed as behavioral norms by human beings in different cultures down millennia of history. Therefore, experimental research on the subject among people of diverse sects, statuses, and nations is likely to yield much fruit. But is this the case for models of explanation such as the D-N model? I believe it is not. True, the concept of causality (which some philosophers deem identical to the concept of explanation) has enjoyed a long and interesting series of experiments (see a brief review in chapter 1 of Michotte's experiments published in 1946). But no less interesting is that, to the best of my knowledge, not one of the creators of explanatory models (reviewed in chapters 2 and 3) has referred to Michotte's studies, and even Knobe and Nichols in their *Experimental Philosophy* do not discuss it (the only place where I found a footnote on Michotte is Psillos, 2002, p. 299).

Beyond noting this matter, whose implications I don't wish to discuss, I'll say just this: I believe there is an important difference between research into moral responsibility and research into models of explanation; empirical research on the former is possible and desirable, but it is hard to argue how it is possible to derive empirical implications about the D-N model itself for the reasons detailed above. (In short, explanation models are procedures, and as such they make recommendations about what should be done, and they are not amenable to an empirical test like an ordinary scientific hypothesis; these procedures are amenable to assessment by justified reasons and the extent of their match with different areas of research.)

PROPERTIES, CHARACTERISTICS, AND CONSTRAINTS FOR CONSTRUCTING A MODEL OF EXPLANATION

At the end of the chapter by Psillos (2007) on models of explanation, he reaches a gloomy conclusion about the concept of explanation:

> In the light of the preceding discussion . . . it should be obvious that there is no consensus on what explanation is. Perhaps, the very task of explaining explanation, if by that we mean the advancement of a single account of what explanation is, is futile and ill-conceived. Perhaps, *explanation* is a loose concept that applies to many things: it is such that it can be partially captured by different models and accounts. (pp. 170–171)

This conclusion, which accords with my own at the end of chapter 2, led me to this question: Based on the fundamental ideas of explanatory models reviewed above, is it possible to attain a (partial) list of properties, characteristics, and constraints linked to giving explanations, a list on which I will be able to base a new proposal for an explanatory model? The present section is devoted to an initial effort to offer an answer to this question. This list contains, in essence, the way I grasp the notion of explanation and how explanations should be given in science. That is, the list constitutes guidelines toward a proposal to construct a model of explanation for psychology (which will be developed in the following chapters).

A List of Constraints, Characteristics, and Properties (the CCP List) for Explanation

CCP 1: It can be posited that what is responsible for the occurrences of the observation under study is some Unknown-Real Process (URP: factor[s], mechanism[s], process[es], and the like) by means of which it is possible answer the questions of why and how the given observation occurred.

CCP 2: The answer to *why/how* questions rests on a theoretical structure (a hypothesis, a model, a theory) that attempts to describe the URP—to approximate it.

Underlying some of the explanatory models reviewed above, some kind of structure was posited as existing in the world on which explanation depended. For example, Salmon (1984) proposed that the world rests on a causal structure, so he characterized a scientific explanation thus: "To give scientific explanation is to show how events . . . fit into the causal structure of the world" (p. 19). He notes that in fact, for Hempel's statement that all scientific explanations strive to propose systematic understanding by fitting into a nomic nexus, he substituted the statement "by showing *how* they fit into a *causal* nexus" (p. 19).

Salmon distinguishes three conceptions of explanation. The first, the epistemic, in principle anchored to Hempel's approach, posits that the given observation will be explained by a logical argument showing that the observation was expected to occur because of the information that appears in the argument's premises (the law or theory and the antecedent conditions). This explanation is characterized as nomic expectability. The second, the modal conception, differs from the foregoing in that it assumes that there is a necessary connection between the *explanans* (the law and the conditions) and the studied observation—that is, given the *explanans*, it is not possible that the observation would not have occurred. This explanation is characterized as a nomological necessity. While the first conception assumes logical necessity, the second assumes necessity as a property of a natural law. The third conception, the ontic, which Salmon proposed and developed, assumes that the world has an elaborate causal structure, and that proposing an explanation is slotting the given observation into the right place in this structure of the world. Accordingly, it rests on causality, on the world's causal structure.

But I hesitate to undertake the commitment that the world has a causal structure or that deduction is a suitable description of processes in the world. My assumption is something more general and abstract, which holds that the occurrence of a given observation is connected to the URP responsible for it. The explanation for its occurrence is handled by theoretical structures (e.g., hypotheses, models, mechanisms, theories) that are relevant to the observation under study, that seek to approximate the general concept of URP, and answer the questions of why and how the observation occurred. I shall call this approach *methodological realism*, which on the one hand assumes that the URP is real, and on the other develops a model that approximates URP and has to be tested empirically, by comparison of predictions with reality.

From this viewpoint, Salmon's ontological position can be regarded as an addition to the assumption about the URP; he suggested a causal characterization of it: the studied phenomenon occurs causally. Similarly, Hempel's epistemic conception can be understood as a description of the URP's behavior by logical relations (deductive or inductive). In other words, Hempel's models can be grasped as an attempt to describe how the URP is connected to the observation under study in terms of logical arguments.

CCP 3: The question, "Does the proposed theoretical structure indeed approximate the URP?" is answered by appeal to the methodological rules of science, the empirical test. When the proposed theoretical structure does pass the empirical test, it is likely to offer an explanation for the observation under study—that is, this structure is likely to answer the questions of why and how the studied phenomenon occurred.

Methodological realism, which is similar to the realist approach, differs from both the instrumentalist approach and the constructive empiricist approach of van Fraassen (1980). By the realist approach, the theories in science describe the events in the world and attempt to present true arguments about it; by the instrumentalist approach the theories are no more than a tool to make successful predictions only, and they do not pretend to describe reality. van Fraassen proposes something different. He does not know if the unobservables that appear in scientific theory indeed describe reality: he is agnostic on this matter. While the realist approach proposes that scientific research reveals what takes place in nature, van Fraassen proposes something entirely different: here is what he writes about constructive empiricism: "I use the adjective 'constructive' to indicate my view that scientific activity is one of construction rather than discovery: construction of models that must be adequate to the phenomena, and not discovery of truth concerning the unobservable" (1980, p. 5).

So methodological realism differs from van Fraassen's approach in that it assumes a realistic approach connected to the URP and to theories in science (which attempt to approximate the URP). By my approach, the goal of theories in science is to discover reality (namely, to reveal the URP), and the way to do it is by constructing hypotheses, models, and theories, whose proximity to the URP is measured by the methodology of science (essentially, the assessment is made by means of the empirical test). There is a small difference between realism and methodological realism: realism posits that the theory reveals the reality, and therefore, it passes the empirical test; methodological realism posits that passing the empirical test supports the theory as approximating URP. In contrast to these, constructive empiricism posits that passing the empirical test is everything there is.

CCP 4: Every explanation has to allow its empirical testing. An explanation that does not allow this is not a part of the scientific approach.

CCP 5: Every explanation is provisional, and as a result of the continuation of research, it is likely to be changed or refuted.

These two CCPs (4 and 5) clearly express the methodological realist approach, whereby the explanation is impossible to detach from the rest of the rules of research methodology, especially the method of empirical testing (the H-D method), which handles theoretical structures (hypotheses, models, mechanisms, and theories) by which means the researcher proposes explanations for the observations under study. In this regard, I follow in the footsteps of Hempel (1965): "The explanans must have empirical content; i.e., it must be capable at least in principle, of test by experiment or observation" (p. 248). These two CCPs also express the spirit of the approach by Popper (1995) that a scientist's work in a given field does not end, and hypotheses unable to satisfy the requirement of an empirical test are ad hoc and are rejected by science (see also Einstein & Infeld, 1938/1950).

CCP 6: When an observation is explained by a theory (which has not yet been refuted), it is posited that this theory is conceived as provisionally correct. It is not desirable to propose an explanation from a false theory, even though this theory may in several cases suggest explanations that hit the mark. When a theory is tested empirically, it is posited that its epistemic status is neutral (whether true or false).

CCP 6 expresses the methodological game rules for the epistemological status of a theory that supplies an explanation as against a condition in which it is tested empirically. According to this CCP, it is recommended that when an explanation is given, the theory (still not refuted) will be taken as true and not false. Although the false theory is likely to posit precise predictions in certain conditions, it is undesirable that the explanation for the phenomenon under study be given by a false theory. If the theory giving the explanation is known to be false, then it is clear (trivially) that it will be refuted in some other cases, and will be ejected from the pool of scientific knowledge (except perhaps for a historical discussion of the subject). In other words, if the goal is that the theory posits correct explanations for a variety of phenomena, this goal will never be achieved by a false theory. By contrast, the present CCP proposes that when a theory is tested, it has to be positioned in a neutral epistemological status, because if it is known that the theory is false (or true), testing it is pointless, for its epistemological status is already known.

CCP 7: The explanation for the observation under study is given by information that is not identical but is relevant to the observation. In science this information appears in the form of a hypothesis, a model, or a theory, by which means an explanation for the given observation is provided.

CCP 8: The explanation is not given on the level of the observation but on a level that represents the observation, namely, on the level of the language of the representation. This language must possess at least the following two properties: ability to represent the observations through symbols; ability to operate these symbols by means of rational rules, so that from a given series of symbols it will be possible to create systematically a different series of symbols. The explanation is based on a rational description of a connection between the changes in the explanatory symbols (the *explanans*) and the changes in the symbols to be explained (the *explanandum*).

CCP 9: For the explanation on the representational level to match reality, there has to be a very close match between the theoretical concepts (on the representational level) and the observations (which describe the reality). Absent such a match, the explanation is liable to miss its mark.

CCPs 7–9 express methodological game rules for the theoretical level on which the explanation is given, the language in which it is given, the connection between the *explanans* and the *explanandum*, and the connection between the theoretical level and the observational level (namely, the observations that it is intended to explain). The explanation itself is not given on the level of reality itself but on a level that reflects reality, that is, on the level on which some theoretical structure is proposed whose purpose is to reflect the URP and the connections between it and the observation (the phenomenon) under study. (That is, changes in the observation are explained rationally by changes in the *explanans*.) The URP is not identical to the observation itself, because we are not prepared to accept that the observation will explain its own occurrence. In fact, when an explanation is asked for, different information is requested from the information under study (the observation under study), but this information must be relevant to the subject. My concern is not to nail down the term *relevance*, so I shall present two illustrations that substantiate what I mean by the notion of relevance. Considering present knowledge, no one will try to explain the free fall of bodies by an appeal to psychoanalytic theory; and the reverse: no one will explain Mr. Aaron's suicide with a bullet fired from a handgun into his temple by detailing the handgun's manufacturing process. In both cases, it is clear at once that these theories are not relevant.

For the explanation to reflect reality (the unknown process responsible for the occurrence of the observation under study in a given condition), we have to represent the URP and its ties to the studied observation by means of a language that allows rational transformations on its symbols. (These transformations appear, for example, in deductive, inductive, mathematical, probabilistic language, computer languages, and practical considerations. In the computer, the output states are generated by certain rules, programs that operate on the input states.) These are languages that allow systematically the creation from one series of symbols another series of symbols. If these transformations were not systematic, it would not be possible to propose an empirical explanation, because the explanation posits that the world is orderly and the events in it are not due to chance or chaos.

As everyone knows, the language of mathematics has proven the most suitable for the most successful describing and explaining of phenomena in nature. I do not mean by this that if we take any mathematical expression it will necessarily be suitable for the explanation of the given observations. What I mean is that mathematics language, with all its complex elaborations, describes and explains empirical phenomena with astounding success. The question arising is why/how does this thing happen? The riveting book by Livio (2009) titled *Is God a Mathematician?* discusses inter alia the present question (the question was raised by Eugene Wigner, Nobel laureate in physics, who wrote an article about the incredible efficacy of mathematics in the natural sciences). Livio summarizes the relevant literature and looks at several possible answers to this question, which may be condensed as follows (and here I do not intend to discuss the question of whether math is a kind of invention or discovery). The amazing success of mathematics may well have to do with the fact that

1. scientists in advance (and also after the fact) match the theory expressed mathematically to the possible empirical results of their research. In other words, we look at the world through the spectacles of mathematics;
2. it is possible that the world itself is nothing but a kind of realization of mathematical relations;

3. because of evolutionary processes, humans have several innate mathematical abilities that foster their extremely efficient survival with the help of the approach that the world behaves mathematically;
4. our impression that mathematics is so successful is nothing but an illusion. In support of this claim, the fact is stressed that in many research areas the success of mathematical models is mediocre or worse, for example, models for weather forecasting, economic models, and of course the inefficiency of mathematical models in psychology.

Although these answers resolve the mystery a little, the question still remains: Why is the mathematical tool so suitable for dealing with a variety of phenomena in the world? What are the properties of this instrument that make it predominant in scientific research, one without which it would be impossible to understand the impressive developments in the sciences? As a hint at the answer, I would like to highlight briefly the following property. As I emphasized in chapter 1, in physics measurement units have been found (such as distance, weight, time) that uphold empirically certain mathematical properties (such as transitivity and additivity), so that what is stated mathematically/theoretically is upheld empirically. I do not know of any other language that sustains so close a proximity between its theoretical, symbolic concepts and reality.

CCP 10: In CCP 8 I posited that the connection, the relation, between the *explanans* and the *explanandum* is a rational one. Here, I shall try to classify generally the kinds of rational relations between the *explanans* and the *explanandum.*

Based on chapters 2 and 3, I shall try to review several rational connections between the *explanans* and the *explanandum* on which researchers have based their explanations. I posit that all these connections attempt to reflect certain processes that take place in nature. I assume that these relations are rational in the sense that they are not an expression of chance and chaos, but a reflection of the deep insight that the world behaves in an orderly fashion.

a. *Relations of argument.* Deductive or inductive arguments. The *explanandum* derives from the premises of the argument by rules of inference.
b. *Mathematical relations including probability and statistics.* The *explanandum* is a certain function of the *explanans.*
c. *Causal relations.* The *explanandum* constitutes a result of the *explanans/cause.* The factor that explains finds expression in various ways: as a necessary or sufficient condition, as a physical causal process, as a factor subject to change by manipulation, as counterfactual conditions.

d. *Relations expressed by a mechanism.* The *explanans* is decomposed into its parts (where each part has a certain function), which are arranged in a certain way that allows the whole mechanism to produce the studied behavior (i.e., the *explanandum*).
e. *Relations based on practical considerations.* The *explanandum* is an action that one performs on the basis of rational/practical conscious considerations that determine that this act will realize one's will.
f. *Relations of inclusion.* It is possible to suggest that in psychology, empirical generalizations of different orders exist, such that the generalization of the highest order is likely to be an explanation of the generalization of a low order. The generalization of the lowest order (generalization 1) is a simple observation that states: *X* (David) responds with response *R* to stimulus *S* at time *T* in place *P*. This observation can be empirically generalizable beyond: subjects (humans, animals), responses, stimuli, times, and places (here we have five dimensions in all). So a simple observation, generalization 1, is explicable by generalization 2 (a group of people responded with response *R* to stimulus *S* at time *T* and at place *P*). Generalization 2 is explicable by generalization 3, which is generalization 2 plus a generalization beyond the dimension of time; and generalization 3 is explicable by generalization 4, which is generalization 3 plus a generalization beyond place. And so on. It is reasonable to suppose that in psychology, a generalization beyond all these five dimensions will be considered a law. (And another important point: these generalizations uphold the rule of transitivity—namely, if generalization 3 explains generalization 2, which explains generalization 1, then generalization 3 also explains generalization 1, simply by inclusion: generalization 3 includes generalization 2, which includes generalization 1.)

CCP 11: Transitivity of explanation is not upheld in all cases.

In brief, if A explains B and B explains C, not in all cases will it be possible to suggest that A will explain C. Transitivity is preserved when the requirement to explain C by A is equal to the requirement to explain B by A and C by B. If these requirements differ, transitivity is not preserved. For example, the requirement to fire the bullet from the barrel of the gun is realized by a powerful and rapid expansion of gases (produced by the ignition of powder in the cartridge case), while the requirement to cause the expansion of the gases is realized by the strike of the pin (a mechanical action), which ignites the powder in the cartridge case. Therefore, the strike itself cannot eject the bullet in all cases, and transitivity is not preserved in all cases. Here is another example. A is the mother of B and B is the mother of C; clearly A is not the mother of C. Nevertheless, the following two possibilities cast a doubt on the nontransitivity conclusion: (a) suppose we change the concept of motherhood to include a woman who takes care of the baby from the day baby is born. (b) Suppose B's inseminated eggs are planted in A's womb. Would not we change the nontransitivity conclusion under these two conditions?

CCP 12: Awareness of the *explanans*, of the *explanandum*, and of the explanation process is understanding the explanation itself in all its components.

CCP 13: It is not possible to suggest for the *explanandum*, for the nonunderstood observation, an explanation by the *explanans*, which itself is not understood. The level of understanding of the *explanans* must be higher than that of the *explanandum*.

CCP 14: Explanations based on models of explanation in the sciences can be obtained by a computer (or a sophisticated robot). By contrast, the teleological explanation (when *X* *wants* to attain *G* and *believes* that action *A* will realize his will, *X* will perform *A*—an explanation we shall denote [Will/Belief]) cannot be given by a computer because this explanation is founded on the individual's consciousness/understanding, and a computer is not gifted with this wonderful quality.

CCPs 12–14 express methodological game rules about connections between understanding and explanation, which for the most part I discussed in chapter 1. CCP 12 highlights the impression that awareness/understanding are connected to a network of contents of the explanatory process (including of being conscious of awareness/understanding). CCP 13 highlights the intuitive idea that the relevant information for an explanation must be on a theoretical/empirical intelligibility level higher than that on which the studied observation is focused. Otherwise the explanatory contribution to the studied phenomenon will be nil.

I would like to emphasize that these three CCPs, especially number 14, show up the sharp difference between the structure of explanation accepted in the sciences and the structure of explanations for the behaviors of human beings (and of higher animals such as primates, dolphins, dogs, and cats). The basic difference between explanations in science and explanations in the social sciences, which I shall call the Conscious Distinction, is the following:

> An explanation in the natural sciences does not appeal to consciousness—it is mechanistic. By contrast, human behavior is interwoven with consciousness. Therefore, a mechanistic explanation of human behavior (which can also be given by a computer) is not complete.

To clarify this phrase, Conscious Distinction, we may look at the following illustration. Assume that David wishes to invest a certain sum of money for profit. He consults with the Financial Advisor software, and after the computer gets David's answers to some questions, David is presented with two options. With option A, he has to invest $10,000, but this way of investment carries many risks, so that the chances of making a profit are 0.001; with option B, he has to invest $100,000, but here too there are risks, and the chances of making a profit are 0.01. What is he to do? David is in a quandary, because it is easy for him to get $10,000, but hard to gather $100,000. The Financial Advisor program's suggestion is pure and simple: objectively, the expected profit from option A is $10, while from option B it is $1,000, so you must choose option B. And now the question arises, what will David do? If he struggles to get the $100,000, we may say that Financial Advisor predicted his behavior and explained it rationally. But if David behaves otherwise—for example, his interest in financial investments suddenly fades—Financial Advisor will shrug its shoulders and say, "Alright, David is not a rational person!"

The important point here is this: it is not hard to program a model of explanation as long as the model deals with objective aspects of behavior. The explanatory story changes dramatically when the individual's behavior has to be explained, taking into account his/her awareness/understanding. To consolidate this interpretation, consider the following illustration, the [Will/Belief] explanation. David wants to hear the heavenly adagietto in Mahler's Fifth Symphony, and believes that a trip to the Israel Philharmonic Orchestra conducted by Zubin Mehta in Tel Aviv will realize his will, so he drives in his car at the appropriate time to Tel Aviv. The journey to Tel Aviv, therefore, is explained by the present account. The important components in this explanation are: David's will and belief; mental contents located in David's consciousness; the will that awoke in David—for example, after he saw an advertisement about the performance of the Fifth Symphony, and believed by virtue of a rational consideration that he must drive in his car from his residence to the concert hall in Tel Aviv, and that this journey was in fact the best way to realize his will out of alternative ways. A computer/robot is incapable of creating such an explanation, because it is not graced with conscious understanding of will and belief. (This argument presupposes that consciousness has not yet been accounted for by neurophysiological processes, i.e., that a mind/body theory has not yet been created. The foundation of this is provided in the following chapters.)

Can David's present behavior be explained by recourse to the models reviewed previously? I think not. For example, it is hard to find any law, theory, or stable empirical generalization by whose means we may explain David's journey. Not only are there no theoretical structures of this kind, but also performing the action that will realize David's will is connected to understanding and following traffic rules invented by human beings living in the state of Israel. It will also be hard to explain his behavior on the basis of causality, because the causality on which the reviewed models rest is physical/neurophysiological, and no theory has as yet been found that explains how consciousness/understanding is constructed or grows up on the basis of the neurophysiology of the brain (see Rakover 2011/2012a, 2011/2012b).

CCP 15: A model of explanation is an integral part of the methodology of scientific research, whose aim, among other things, is to provide understanding, that is, answers to the questions of why and how the given phenomenon occurred. In this respect, it is possible to grasp an explanatory model from three methodological viewpoints:

a. *Procedure:* A model can be viewed as a procedure, a series of instructions that the scientist must follow to propose an explanation acceptable methodologically for the given phenomenon.
b. *Justification:* A model is based on several theoretical arguments that constitute justifications for why the model is to be understood as a procedure for giving scientific explanations.
c. *Theory:* In certain cases, a model is founded on a theory with specific content (scientific, philosophical such as causality) that justifies the evaluation that the model is scientific.

The third viewpoint (c) seems to me very problematic, and in my opinion it is the one that has drawn the most fire in the criticism of the models reviewed above. The reason for this stems from the lack of distinction as to what is the function of a model of explanation: to propose (a) a theoretical approach to scientific explanation or (b) a procedure for giving scientific explanations. To clarify this distinction, I shall very briefly analyze three illustrations that are interpreted on the lines of the present distinction. The first is Bomberger's criticism of the D-N model regarding the case of the flagpole and the shadow. This criticism is not aimed at the model's procedure and justification (according to which the prediction was expected because of the information presented in the premises of the model) but at the theory (the law) that appeared in the premises. Had the law/theory also included a clear causal indication for the fact that the flagpole cast a shadow, and so on, the critical argument that the present model does not provide an explanation (causal) but only a prediction would have been eliminated. (Hempel's suggestion to distinguish successive laws from simultaneous laws as a way to handle Bromberger's criticism may be seen as support for the present position.)

The second illustration is that of the barometer. To my mind, this criticism is intended to indicate a need for a series of experiments to show that, indeed, the factor common to the fall in the barometer and to the appearance of the storm is the steep fall of barometric pressure.

The third criticism is that directed at Salmon's C-M model in respect of the chalk spot. This critique does not negate the model itself but Salmon's causality theory, because it is hard for the latter to distinguish a relevant from a nonrelevant cause. But this criticism can be overcome by one or two experiments to invalidate the explanation of the chalk spot. For example, the same strike at the same angle with a cue whose tip has not been chalked will give rise to the same observations on the ball and the same measurements as in the case where the chalk spot is transferred to the balls (one may reasonably assume that the effect of the chalk spot itself in physical terms is negligible).

What arises from this short discussion is this: when a model of explanation is under discussion, one must concentrate on the first two viewpoints, and understand that the explanatory information (the theory, the law, the stable generalization) is not always capable of answering all possible questions, and that often the answers to these questions are given in the framework of further research.

An Explanatory Procedure for Psychology

Where do all these reflections lead us? They have presented me with the following outline of an explanatory procedure for psychology; it is to be based on the following three major theoretical components:

1. *Explanatory information:* The procedure should include relevant information in the form of theoretical structures: hypotheses, models, theories, stable generalizations, mechanisms, and the like. This information should be different from the information describing the studied behavior (phenomenon), and it should lead to answers to *why/how* questions regarding the behavior occurrence. It should approximate the URP, which is assumed to be the real process responsible for the actual occurrence of the behavior under study.
2. *Antecedent conditions:* The procedure should include a description of the appropriate situation (e.g., stimuli) to which the studied behavior is connected. These conditions are vital for giving answers to the *why/how* questions.
3. *Rational relations:* The procedure should employ certain rational relations that connect the antecedent conditions to the explanatory information (i.e., the theoretical structures) in such a way as to produce a particular prediction, which is supposed to be identical to the description of the studied behavior. These rational relations can take the form of logic (deductive, inductive arguments), mathematics, statistics, practical considerations, and the like.

As can be seen from this sketch, the proposed procedure is a kind of a generalization abstracted from all the models reviewed above (see chapters 2 and 3). For example, the appropriate place of the natural law in the D-N model has been widened to include various theoretical structures, and the deductive relation of this model has been widened to include all sorts of rational relations. In addition to these, the proposed procedure includes the most important concept of URP. In my view, this concept is covertly assumed by almost all the explanatory models reviewed above: it is the latent assumption that in the real world something is responsible for the occurrence of the behavior (phenomenon) under study, an assumption that I named the Unknown-Real Process (URP).

The proposed procedure consists of the following three theoretical levels:

1. *Theoretical-explanation level:* consists of the theoretical structures, which are supposed to approximate the URP on theoretical level (II);
2. *Theoretical level (II):* consists of the assumed URP, which is responsible for the real occurrence of the studied behavior; and
3. *Observational level (I):* consists of the description of the behavior under study.

The connections among the concepts of these levels (and among those within a level) are made through the use of rational relations. (A detailed development of these ideas is set out in chapter 7.)

Chapter 5

Methodological Dualism (MD) and the Multi-Explanation Framework (MEF) (I)

Theoretical and Empirical Basis

Chapters 5 and 6 set out in two parts my approach to explanation of behavior in psychology (see Rakover, 1997, 2007, 2011/2012a, 2011/2012b). The MD (Methodological Dualism) stands for arguments showing that a mentalistic procedure (model, schema) to propose teleological explanations based on the individual's will/belief meets the methodological requirements of science even though it cannot be reduced to concepts prevalent in the sciences; the MEF (Multi-Explanation Framework) shows how it is possible to construct a specific and coherent theory in a certain research domain that is founded on two kinds of explanatory procedures: mechanistic (based on concepts prevalent in the sciences) and mentalistic (based on concepts concerned with the individual's inner world such as her will, belief, intention, and consciousness). Chapter 5 prepares the methodological ground for developing the MD and the MEF. It shows that to date we have not succeeded in conceptualizing or converting the subjective concepts referring to the individual's inner world into objective concepts prevalent in the sciences. In chapter 6 I shall present and justify the MD and the MEF approach.

Based on the preceding chapters, one may have the impression that researchers have designed various models for a scientific explanation according to their various research domains. For example, the explanatory model for deterministic phenomena is clearly not applicable to statistical phenomena. Thus, the D-N model, which is highly suitable for phenomena that describe the movement of bodies handled by Newtonian theory, is not suitable for statistical phenomena; therefore, Hempel proposed models suited to this kind of empirical observation (the D-S and the I-S models)—likewise Salmon (the S-R model). While chapter 2 reviewed the explanatory models tailored to research fields in physics and biology, chapter 3 presented three important models more appropriate for understanding human behavior (and the behavior of supreme animals). But even these three kinds of models—the rule-explanation model, the teleological model (functionalist), and the new-mechanistic model—cannot, in my opinion, contend with the extremely important and fundamental factor in explaining human behavior, namely, the experiential-awareness factor, otherwise named the conscious-experience factor (below I use both names interchangeably). Absent this factor, it is not possible to understand human behavior, or animal. Very briefly, my approach suggests that to propose a better explanation for human behavior than that based solely on a mechanistic explanation, there also must be added in a coherent manner a mentalistic explanation. As may be understood, the name Multi-Explanation Framework expresses the methodological approach whereby an explanation of behavior must be based on two kinds of procedures: mechanistic and mentalistic.

This approach raises the following crucial question: can experiential awareness, the behavior interwoven with conscious experience, be captured by means of the mechanistic concepts prevalent the sciences? If the answer is affirmative, if it were possible to understand the conscious experience in terms of the natural sciences, the methodological basis for developing my approach: the MD and MEF (which employs both kinds of explanations—mechanistic and mentalistic), would be demolished; everything could be explained by means of the conceptual framework developed in the natural sciences. McCauley and Bechtel (2001) wrote along similar lines in the context of reducing a psychological theory to a neurophysiological theory: "If psychological theories map neatly onto neuroscientific theories along the lines that the classical model specifies, then our commitments to psychological states and events are, at the very least, dispensable in principle" (p. 739).

But this is not the situation at present. In this chapter I review in brief several important approaches found in the professional literature that attempt to propose different ways to conceptualize and capture mentalistic concepts by means of mechanistic concepts. This effort has not borne fruit (such is my impression); therefore, the following conclusion may be suggested:

We still have no mind/body theory detailing the mechanism whereby neurophysiological activity in the brain creates or acquires consciousness, akin to the accepted theories in science: physical theories that explain the transformations of energy (associated with potential and kinetic energy, friction and heat, magnetism and electricity, mass and energy) or how matter changes from one kind to another, like the theory of how hydrogen and oxygen join to form water, and how water can be broken down (by electrolysis) into these gases.

Several researchers and philosophers reached a similar conclusion about the nonexistence of mind/body theory:

1. "Whatever our mental functioning may be, there seems to be no serious reason to believe that it is explainable by our physics and chemistry." (Putnam, 1975b, p. 297)
2. "We have been trying for a long time to solve the mind-body problem. It has stubbornly resisted our best efforts. The mystery persists. I think the time has come to admit candidly that we cannot resolve the mystery." (McGinn, 1989, p. 349)
3. "The sensation of color cannot be accounted for by the physicist's objective picture of light-waves. Could the physiologist account for it, if he had fuller knowledge than he has of the processes in the retina and the nervous processes set up by them in the optical nerve bundles and the brain? I do not think so." (Scrödinger, 1992, p. 154)
4. "To be brutally honest, scientists do not yet have even the remotest idea of how visual experiences—or indeed any other kinds of experiences—arise from physical events in the brain." (Palmer, 1999, p. 618)
5. "The reason the mind-body problem does not go away, despite our being clear about the options in responding to it, is because of the constant battle between common sense, which favors the view that the mental is a basic feature of reality, and the pull to see it as an authoritative deliverance of science that this is not so. We find ourselves constantly pulled between these two poles, unable to see our minds as nothing over and above the physical, unwilling to see the universe as containing anything not explicable in terms of its basic, apparently non-mental, constituents." (Ludwig, 2003, pp. 29–31)
6. "Even if we accept the familiar idea that minds are somehow dependent on brains, we have no clear idea of the nature of this dependence. The mental-physical relation appears utterly mysterious." (Heil, 2003, p. 217)
7. "The problem of consciousness is completely intractable. We will never understand consciousness in the deeply satisfying way we've come to expect from our sciences." (Dietrich & Hardcastle, 2005, p. 1 [the opening sentence])
8. "Are neuroreductionist explanations of cognition possible?" (This is the title of Uttal's [2014] paper, to which he proposed a "probably not" answer for methodological, conceptual, and empirical reasons [p. 37].)

In drawing up this list, I feel duty-bound to add the following rider: Colin McGinn is regarded as one of the most important researchers in the group promoting what Flanagan (1991) termed "the new mysterianism." This group's approach holds that the problem of understanding, the explanation of the conscious experience, is inexplicable, and apparently exceeds human intellectual capacity. Although McGinn is numbered among the above eight researchers in whose opinion a mind/body theory has still not been discovered, to the best of my knowledge, it will be hard to include the other researchers (including myself, of course!) in the new mysterianism group. (I stress yet again that when I speak of the nonexistence of a mind/body theory, I highlight the notion of time: not yet. As to whether such a theory is likely to develop in the future, my answer is that I don't know.)

To consolidate my impression, and to bolster these eight statements, I shall try to answer several questions that itemize the global question above: Is it possible to capture conscious experience in the concepts of the natural sciences? These are the questions:

1. Can human behavior be explained mechanistically alone, without recourse to the concept of experiential awareness?
2. Has a theory been successfully developed that explains the connection between neurophysiological processes in the brain and conscious experience?
3. Can mentalistic explanations anchored to the concept of conscious experience be reduced to mechanistic explanations (e.g., to a neurophysiological explanation)?
4. Has a very complex and sophisticated computer been able, like us, to behave with conscious experience?

An affirmative answer to any of these four questions means that there is no need for a special explanatory model to handle human behavior (or the behavior of higher animals), nor to develop MD and MEF, because the standard methodology developed in the sciences definitely accords with research in psychology. Only the negative answer to all these questions obliged me to think about the development of a new explanatory approach suited to the field of psychological research. But before entering into a discussion of these matters, I must offer a clarification of the following basic question: What do I mean by the concept of conscious experience?

The term is firmly tied to a person's inner world, his feelings, emotions, thoughts. Above all, it refers to the wondrous experience in which the person is, to experiential awareness (and I assume that a large proportion of animals are situated at some levels of conscious experience). Actually, I believe that a person's regular day-to-day behavior (feelings, emotions, perception, learning, remembering, social behavior, etc.) is for the most part saturated with experiential awareness. With everything he does, he is in a state of conscious experience, awareness of his actions, of the world he is in, of his body, and of what occurs in his consciousness. This experiential awareness contains the person's entire perceptual world, but concurrently he can center this awareness on one thing or another, according to his needs and requirements. This is a subjective experience of awareness, a state of consciousness that the individual himself alone is aware of and is situated in. The literature customarily calls what I describe "qualia" (e.g., Tye, 2015). Here I shall not discuss the many debates around this term (e.g., do qualia depend on, or are they independent of, neurophysiological processes occurring in the brain?), but I shall stress the following quality that is linked to experiential awareness. I rely on Levine's (1983) article. Levine argues that even if our knowledge of the neurophysiological and chemical action of the brain were truly perfect, our explanation for behavior would still not be complete; something would still be lacking in the explanation, something that the physical/chemical/neurophysiological explanation has not succeeded in reaching—and this is experiential awareness, qualia, the subjective conscious experience of each and every person. To date, no one has succeeded in understanding how neurophysiological (physical, chemical) changes are connected to changes in the subjective experiential awareness of each and every person (and the reverse). Levine calls this "the explanatory gap." (I named it "the nonexistence of a mind/body theory.")

Is experiential awareness subject to error? My answer is negative. I maintain that experiential awareness itself, for all its levels and diverse aspects, is not subject to error. By that, I do not mean to say that a person is not liable to produce a mistaken conclusion or to arrive at a wrong judgment on the basis of her experiential awareness. Clearly, if the individual does not know exactly what the cause is of the awful headache that has overwhelmed her, she is likely to draw mistaken conclusions as to the factor responsible for this sensation of pain; moreover, she may mistakenly think that her headache is the most excruciating of pains, unequalled by any other. But one cannot deny the very sensation of an aching head, the conscious experience that engulfs the person. I shall substantiate my meaning with two illustrations.

The first is the famous illusion of Mueller-Lyer (see Coren & Girgus, 1978), shown in Figure 5.1. The perceptual impression one gets from Figure 5.1 is that the left-hand line is longer than the right-hand line. This is experiential awareness, on the basis of which a person will answer the question, "Which line is longer?" with the reply, "The left-hand line." This judgment, of course, is wrong, as the lines are equal in length. Despite this knowledge, the experiential awareness remains firm. The perceptual conscious experience remains as it is, exactly as the person experienced it before being told that the lines are equal in length. This is not the place to argue with him, and tell him that his conscious experience is wrong, that what you experience is incorrect because

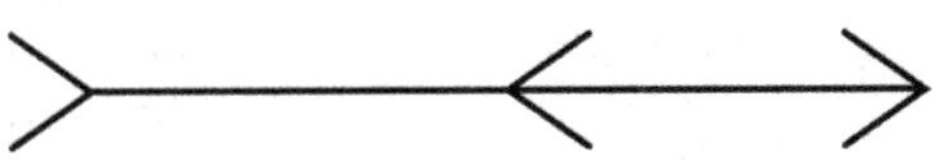

Figure 5.1 The Muller-Lyer's Illusion. The left horizontal line is perceived as longer than the right line although the two are equal.

the lines are the same length. Why? Because despite the knowledge that the lines are equal, consciously the person experiences lines of different size; this is precisely what one experiences, and this subjective conscious experience is not to be disputed.

The second illustration concerns an experience I underwent many years ago. It was when 3-D movies were being screened for the first time in Israel. At the very opening of the three-dimensional film, which depicted battles among wild tribes, a spear was hurled at me from the screen. Because of the three-dimensionality, it seemed to me that the spear was about to strike me on the forehead between my two eyes. Suddenly, at the spot where the hurtling spear was expected to strike, I felt a very sharp pain. The spear, naturally, did not penetrate my brow, but the pain was there, despite my absolute knowledge that the thrust spear was just a part of the movie, not real. This was a conscious experience so clear and sharp that I remember it to this day. The pain passed soon afterward, and never returned, all the arrows and spears flung at me notwithstanding. So, here is experiential awareness, the sensation of conscious pain, sharp and clear, which was not caused by anything real and physical, but by a 3-D illusion. Nevertheless, no one can argue with me and tell me that I was wrong, and that at that moment I did not experience sharp pain in the forehead between my eyes; and despite the fact that since then I have watched dozens of 3-D films, and volleys of bullets and shells have been "fired" at me, and aircraft have flown at great speed and passed me on either side, I have never felt the sensation of pain in my forehead; and, if I am not mistaken, nor have I responded by flinching, but in many cases I expressed my amazement at the technical stunts accomplished in the movie with a cry of "Wow!" and excited laughter.

These two illustrations substantiate the point I wish to emphasize: while the conscious experience itself, the experiential awareness, is not subject to error for the individual himself who undergoes the experience, everything associated with this experience (judgments, conclusions, explanations, and so on) are indeed subject to errors and mistakes of various kinds. In this respect, I agree with Churchland (2013), who does not accept the traditional approach to the concepts of consciousness and introspection: "Accordingly, one's candid introspective judgment about one's own mental states—or about one's *sensations,* anyway—are incorrigible and infallible. It is logically impossible that they be mistaken. The mind knows itself first, in a unique way, and far better than it can ever know the external world" (p. 123).

In my opinion, experiential awareness is an incontrovertible existential experience of the human being (and of supreme animals). Because the very experience of my existence lies in my experiential awareness (which, as stated, is interwoven with all my activities and deeds), it arises that to cast doubt on this experience, that is, to suggest that I am liable to err in the very existence of conscious experience, seems paradoxical to me, because this doubt is also cast on the very experience of my existence, and the casting of doubt itself is clear evidence of my being alive; and the interpretation of my being alive is that I am situated in conscious experience. And this paradox can also be formulated in the Cartesian spirit: casting doubt on experiential awareness is itself no other than existential awareness; therefore, this conscious experience of casting doubt is a kind of negation of itself, of casting doubt.

CAN HUMAN BEHAVIOR BE EXPLAINED MECHANISTICALLY ALONE, WITHOUT RESORTING TO THE CONCEPT OF EXPERIENTIAL AWARENESS?

Thus, after clarifying what I mean by the concept of experiential awareness (qualia, conscious experience), I move on to this important and interesting question: Do we need this concept to explain behavior? Cannot behavioral phenomena be explained without recourse to the concept of experiential awareness?

Several years ago I had a discussion with colleagues at the department of psychology at Haifa University about whether conscious experiences, experiential awareness, qualia, are needed for an explanation of behavior. They maintained that any behavior can be explained mechanistically without resorting to consciousness. I did not agree. I believe that my colleagues were expressing the common approach to this issue (held by behaviorism and cognitive psychology): conscious experience is not needed to explain behavior. Why? Basically, because of the following "otiose" argument: given that many mathematical functions can be fitted to any set of empirical observations, and given that these functions express different mechanistic theories, it follows that a mechanistic theory may be constructed for any set of psychological observations. In other words, given that for any set of observations, many and varied computational programs exist that could generate this group of observations, a theory, founded on computational, neurophysiological components, may be constructed for these observations. Hence, one may ask why we should appeal to the complex hybrid MEF (mechanistic and mentalistic explanations) if we can explain a given phenomenon successfully by appeal to a mechanistic explanation only. That is, consciousness apparently does not enhance our understanding (predicting, explaining) over and above the understanding obtained through a mechanistic theory.

In this regard Dawkins (1995) wrote: "There is no prediction we can make that if the animal has consciousness it should do X but not conscious it should do Y" (p. 139). Flanagan (1992) conceived of "conscious inessentialism" as "the view that for any activity i performed in any cognitive domain d, even if we do i consciously, i can in principle be done nonconsciously" (p. 129). Similarly, I can propose here a relatively new argument, which is based on the *identity theory*, and which runs as follows: given the popular solution to the problem of mental causation, which identifies a mental state (MS) with a neurophysiological state (NS) and which is based on well-understood causal relations among NSs, one may propose that conscious experience is not needed to explain behavior, simply because the complete explanatory job is done by the NSs (for review and debate regarding the *identity theory*, see Robb & Heil, 2014; Yoo, 2015). I don't agree: mental events and processes are essential for the explanation of behavior.

First, as I illustrated in Rakover (2007, 2011/2012a, 2011/2012b), various behaviors do necessitate an appeal to mentalistic explanations, because it seems difficult to propose for them purely mechanistic explanations.

Secondly, I believe that the otiose argument is inadequate. In my view, this argument is based on a crucial hidden assumption, which many researchers have overlooked: the behavior that is explained in psychology is behavior stripped of conscious meaning. The psychological indexes (such as number of correct responses and reaction time) do not carry any conscious meaning since these indexes are the same as those that represent a robot's behavior: they are public responses. Psychologists only treat behavioral properties that belong to the public domain, properties that do not carry private conscious meaning. So if the behavior to be explained is devoid of conscious meaning, no wonder the explanation is constructed mechanistically. But in this case, one does not account for a conscious meaningful behavior but suggests an explanation for a zombie's behavior. Hence, a computer that simulates David's behavior cannot simulate his behavior exactly since it cannot simulate David's meaningful behavior, the behavior interwoven with conscious experience, but only the public behavior, that is, the behavior stripped of consciousness. To demonstrate the point that a computer cannot simulate behavior completely, consider the following question: Can a computer simulate originality (original, innovative creation or behavior)? If the answer is yes, then the computer's imitation is no longer original and innovative. Furthermore, if the computer generates (by employing a special program) a certain invention, how can it judge that its creation is original and innovative? If it relies on a list of previously known original inventions and declares, "My creation belongs to this list, the list of originality," then once again its creation is no longer original. (If the innovation does not belong to the list, it may be evaluated as nonsense.) Hence, the computer cannot imitate all human behavior. If the answer is no, one admits that a computer cannot simulate all human behavior.

Thirdly, the argument from *identity theory* (which suggests that one does not need the MSs to explain behavior, since the whole explanatory job is accomplished successfully and completely by the NSs) rests on shifting sand. Why? Because *identity theory* itself was devastated by severe criticism (e.g., Robb & Heil, 2014; Yoo, 2015). Here is a famous example: in 1982, Jackson proposed his well-known thought experiment, which can be perceived as an argument with negative implications for the attempt to conceive MSs as identical to NSs. Accordingly, Mary, a super-scientist in the area of vision who lived all her life in a wholly black-and-white environment, learned all that can be known (physically, chemically, neurophysiologically, etc.) about color vision. One day, she set out from her black-and-white environment and experienced for the first time in her life a red color. And, so goes this thought experiment, at that moment Mary learned something new, not provided her by scientific knowledge: what it was like to see a red color, the conscious experience of red color. (By the way, this thought experiment has several similarities to Scrödinger's statement above.) This story provoked a wide debate, a review of which is beyond the purpose of this chapter. Nevertheless, my impression from the literature is that identity theory was overshadowed by dark doubts (see Churchland, 2013; Horgan, 1984; Robb & Heil, 2014; Yoo, 2015).

Following Flanagan (1992), who proposed the "conscious inessentialism" approach (see above), according to which a conscious action can be explained by appeal to unconscious processes, I divide research on consciousness into two groups. A number of researchers have developed an approach that I shall call "consciousness-dispensability," which proposes that consciousness has very minor importance in the explanation of behavior (e.g., Bargh, 2007; Bargh & Ferguson, 2000; Bargh & Morsella, 2008; Dijksterhuis & Aarts, 2010; Nisbett & Wilson, 1977; Roser & Gazzaniga, 2004; Umiltà, 2007; Velmans, 1991; Wenger, 2003; Winkielman & Berridge, 2004); by contrast, other researchers have developed an approach, which I shall call "consciousness-indispensability," that proposes that consciousness enjoys weighty importance in explanation (e.g., Baars, 2002; Baumeister, 2008; Baumeister, Vohs, & Funder, 2007; Funder, 2009; Gollwitzer, 1999; Locke & Latham, 2002; McGovern & Baars, 2007; Mitchell, De Houwer, & Lovibond, 2009; Morsella, 2005; Rakover, 1993, 2007; Sun & Franklin, 2007; Weidemann, Satkunarajah, & Lovibond, 2016).

Neither of these approaches seems extreme: the former does not suggest that there is no need at all for consciousness, and the latter does not suggest that all behavior is explained by an appeal to mentalistic schemes only. An extreme approach of consciousness-dispensability actually suggests that consciousness is an epiphenomenon and an illusion, while an extreme approach of consciousness-indispensability denies the existence of processes outside consciousness. These are conclusions that are very difficult to accept. As against the consciousness-indispensability extreme approach, it may be said that many processes in the brain and body located outside consciousness and control are known facts. And as against the consciousness-dispensability extreme approach, it may be wondered, what is the survival value of consciousness that is an epiphenomenon and an illusion? It is hard to understand how this strange property of consciousness, devoid of influence on behavior, has developed in humans and animals. I shall briefly consider a few of these examples.

Although Umiltà (2007) may be considered a supporter of consciousness-dispensability, he reaches the conclusion that we are aware of few aspects in behavior. And although Wenger (2003) suggests that conscious free will is an illusion, he tends not to believe that consciousness is ineffectual. Given these two views, one may suggest that a full explanation of behavior will be made by an appeal to both conscious and unconscious processes.

In 1977 Nisbett and Wilson published a famous article in which they claimed that analysis of many experiments in social psychology and decision-making showed that often participants had no awareness of the stimulus, of the response, of the effect of the stimulus on the response, or of higher cognitive processes. This claim was based on findings that the explanations given by the participants for their own behavior were incorrect.

In response, I suggested that even when the explanations are incorrect, the participants might have access to their cognitive processes. Just like scientists who are liable to put forward incorrect hypotheses based on external observations, so the participants are liable to put forward incorrect explanations based on internal observations (introspection) (see Rakover, 1983).

In an extensive paper, Velmans (1991) argues that in many cases (e.g., speech), awareness of the behavior comes about when it is being performed or—and this is the most frequent—afterward. From examples of this kind, Velmans concluded that consciousness has minor importance in processing information, and that unconscious processes are those that mediate between the stimulus and the response.

In reply to Velmans's paper, I constructed a thought experiment that I called "the mental-pool thought experiment," based on the common assumption that the capacity of consciousness is limited, while the amount of unconscious information is unlimited (Rakover, 1996). On the grounds of this thought experiment, I argued that Velmans's approach was correct only in part: the processing between the stimulus and the response is in part unconscious and in part conscious. At first the information is treated unconsciously, but subsequently some of the information penetrates consciousness, and therefore, an opportunity is created for conscious information to initiate, monitor, and control behavior. Once again, it may be suggested that an appropriate explanation for behavior has to appeal to both conscious and unconscious processes.

Libet and colleagues' experimental findings, which are discussed in many of the above articles, support the consciousness-dispensability approach (e.g., Libet, 1985; Libet, Gleason, Wright, & Pearl, 1983). Libet's main finding is based on the measurement of the times of occurrence of three processes in a participant: first process—a voluntary movement of the wrist (a process marked *M*); second—introspective reporting of the time when the free will to move the wrist arose (*W*); and third—measurement by EEG of the cerebral action that preceded the voluntary action, called readiness-potential (*RP*). The results revealed that *RP* preceded *W* by approximately 350 ms, and *W* preceded *M* by approximately 150 ms.

Without discussing the criticism of this work (see, e.g., open peer commentary and author's response in Libet, 1985) let us consider Libet's main conclusion: the source for the voluntary action is subconscious, but consciousness has control over the action in the time span between *W* and *M*; for example, the individual has the power of veto over the voluntary action. A slightly different interpretation may be suggested as well. Since the movement of the wrist is not a purely motor movement, as it carries meaning (the individual's intention), the two processes *RP* and *W* may be regarded as necessary conditions for *M*: if *RP* does not occur, the motor movement will not be realized, and if *W* does not occur, a meaningless movement will appear. Hence, the explanation of the behavior cannot be mechanistic alone and must also take into account an appeal to the individual's conscious will/belief.

Similarly, Baumeister (2008) distinguished two systems of action: free and unfree action, where the former (characterized by flexible, controlled, and conscious processes) starts to operate when the need arises to navigate and change the behavior.

On the basis of a survey of research in learning without awareness conducted from the 1930s to the 1990s, Rakover (1993) concluded that rule-based learning was accomplished with the help of cognitive processes located in consciousness and sometimes outside it, a conclusion that supports the notion that an appropriate account of behavior has to appeal to both conscious and unconscious processes.

Mitchell, De Houwer, & Lovibond (2009) reviewed the professional literature over the past fifty years as to whether associative learning can occur without awareness. They concluded: "There is, therefore, no automatic mechanism that forms links between mental representations. . . . Thus, learning does not take place outside of awareness; it requires cognitive resources, and it is affected by verbal instructions, rules, and deductive reasoning processes" (p. 198). (This is a target article, and it is followed by critical peer commentaries and the authors' responses.) A recent study in eye-blink conditioning demonstrated that awareness of the connection between the conditioned stimulus and the unconditioned stimulus improved associative learning dramatically (e.g., Weidemann, Satkunarajah, & Lovibond, 2016).

The importance of the conscious process in the explanation of behavior is further reinforced by the findings of Simard and Basmajian (1967). They found that it was possible to train participants, by means of visual and auditory feedback, to control voluntarily the frequency of discharge of a single motor unit (the right tibialis anterior). Baars (2002, prediction 5) uses this example in support of one of the seven predictions connected to the effect of consciousness on behavior.

It may be proposed that automatic behavior, which is generated by overlearning and through much practice and training, shows the following characteristics: it is unaware, does not require supervision, needs minimal effort for processing information, is stereotypical, is accomplished very quickly, and so on (e.g., Logan, 1988; Palmeri, 2002; Schneider, 2009). But often, when automatic behavior fails for some reason (e.g., a typo made by a skilled typist) or when a change occurs in the surroundings (e.g., a diversion on a familiar route), conscious processes of supervision and control go into action (see Schneider, 2009). It is hard to accept that the degree of automation in automatic driving with a manual gearshift is identical to that with an automatic gearshift. Logan (1988) writes: "According to instance theory, automatic processes are used intentionally. Automaticity exploits the autonomy of the retrieval process, harnessing it so that it provides information that is relevant the person's goals" (p. 513).

Similarly, I suggest that consciousness is endowed with the ability to "recruit subsystems" that are needed for achieving one's goal. For example, when David plans to go to the movies, he recruits behavioral systems that are in part automatic and in part conscious: going to the cinema, buying tickets, sitting down in the right seat, and watching the film. A similar thing happens with the pianist, the violinist, and the tennis player, who races across the court after the ball and drives an angled backhand passing shot. In these cases, consciousness does not monitor every single movement. On the contrary, consciousness monitors the "behavioral flow," so that the tennis player knows at once that the shot is in, or out.

Likewise, Gollwitzer (1999) suggested that teleological behavior is explained by goal intentions and their implementation: the conscious determination of the goal and the conditions (where, when, how) in which the goal will be realized. And indeed, when the conditions are realized, people pass from conscious considerations to automatic behavior. And Locke and Latham (2002) posited in their "Goal Setting" theory that to realize their conscious goals, people automatically use the appropriate knowledge and skills that they possess (see also Bratman, 2001; Dijksterhuis & Aarts, 2010).

I conclude this section with a very short review of the behavior of Max the cat, the house cat of the Rakover family. I shall show that it is hard to understand his behavior without appeal to mentalistic concepts (e.g., teleological explanation, will/belief: see Rakover, 2007). (And indeed, I am entirely convinced that cats and dogs are gifted with consciousness, possess intentions and wishes, and believe that a certain action is likely to realize their wishes.)

Max the cat is a Himalayan housecat. His living space is the apartment of the Rakover family, where he has lived uninterruptedly since the age of four months. Over the years I have carried out a large number of observations on his behavioral episodes, which I have interpreted as exhibiting mentalistic behavior (see Rakover, 2007). As an example, let us examine the episode of *scratching the armchair*, which I shall divide into the following behavioral components:

- I (Sam Rakover) am seated in my armchair watching television;
- Max approaches me;
- Max scratches the left-hand edge of the armchair (to the right of the armchair stands a table);
- I turn to him, lift him up, set him on my knees, and pet him (stroking, tickling, and so forth).

The interpretation centers on the main behavioral component of this episode, the scratching, a response that starts off the social interaction between the cat and myself, which realizes the cat's will and which requires a teleological mentalistic explanation (for support of the hypothesis that animals have consciousness and will, see, e.g., Edelman & Seth, 2009; Rakover, 2007). The immediate question is how can one interpret Max's behavior? I propose to account for it by an appeal to a teleological explanation, which in itself raises the following question: How does the scratching response fit a teleological explanation? The answer is that the scratching response has undergone a change from realization of the function of survival–adaptation to realization of a new goal: being petted by Sam, a behavior that is accountable in terms of a teleological explanation (called Will/Belief). Max has recruited the scratching response to achieve a new goal. The scratching response in itself is a survival–adaptive response: the cat unsheathes his claws by stretching his paws, when he is on the hunt or in defense–attack situations, in order to assist claw replacement, and to mark the scratched place with his scent. In the present case, the shift from survival–adaptation to getting petted has been accomplished through a learning process. Max has learned that scratching the armchair where I am sitting leads to positive reinforcements: I pick him up, set him down on my knees, and pet him.

This interpretation may be summarized in three stages:

1. Determining the general framework of the explanation: The teleological explanation—Max wants to be petted by Sam and believes (by virtue of

previous learning, acquisition of knowledge) that scratching the armchair will achieve petting, so Max scratches the armchair.
2. Explanation of the scratching response itself: This response in itself is commonly explained mechanistically, that is, in terms of the claws' physiological–anatomical structure and their evolutionary functions.
3. Integration of the scratching response with the teleological explanation: Max has learned that under certain conditions the scratching response leads to realization of his goal: petting by Sam. And now, in many cases, when the desire arises in him to be petted by Sam, he uses this learned knowledge to achieve his goal. This goal (petting) is entirely different from the previous functions achieved by scratching.

Since this interpretation includes a learning element, we may ask: Can this episode be explained as simple instrumental learning (for which a mechanistic explanation can be proposed) of the kind: stimulus → response → reinforcement? The answer is negative, for the following reasons: Max's behavior is not stereotypic. He does not scratch the armchair each time I sit down in it and watch television. Moreover, he continues to scratch all the other chairs and armchairs in the room (regardless of the scolding), and sometimes he jumps onto my knees even without scratching the armchair before doing so. So this episode cannot be treated as simple learning where a given stimulus produces the learned response over and over again. Max is the one who decides when and how to realize his intentions; it is he who initiates this episode, the interaction between us. It is reasonable to assume that Max uses this acquired knowledge (scratching → being picked up, placed on knees, and being petted) whenever he has the will to be petted by me. Similarly, we go to a restaurant when we are hungry and use acquired knowledge (such as the location of the restaurant, the kinds of food served) to satisfy our will.

Another possible simple learning explanation is this: Max has learned that he can scratch the armchair without Sam scolding him for ruining it, so he scratches its left-hand side—and that's all he does! In that case, it is Sam who thinks that Max wants to be petted (even though all that Max does is scratch), and it is Sam who lifts him onto his knees and pets him. Is this interpretation reasonable? In my opinion the answer is negative, because it is not consistent with the following observations.

First, when Max is placed on knees without his request, he tries to get off and run away. By contrast, it has never happened that Max tried to escape from knees in the *scratching the armchair* episode.

Secondly, when Max wants to knead my stomach (an instinctive response connected with suckling), he leaps straight up onto the left armrest of the armchair, from where he moves onto my knees, and while sitting on his rump, he kneads my stomach with his forepaws. When I try to stop the kneading and to sit him or lay him on my knees, he jumps off.

Thirdly, sometimes when I am sitting gazing at the television, and my left arm rests on the chair's armrest, Max taps my arm with his paw, and when I turn my head to him, he sometimes begins to scratch the armchair; sometimes he begins climbing onto the armchair, and sometimes he sits there and waits for me to lift him onto my knees. It is, therefore, reasonable to assume that in this case, as in the *scratching the armchair* episode, it is Max who initiates the interaction between us, sometimes by scratching the armchair and sometimes by tapping my arm.

HAS A THEORY BEEN SUCCESSFULLY DEVELOPED THAT EXPLAINS THE CONNECTION BETWEEN NEUROPHYSIOLOGICAL PROCESSES IN THE BRAIN AND CONSCIOUS EXPERIENCE?

Several researchers suggest that wide-ranging activity of a neural network that unites various functions in the brain (such as the "global neural workspace") is a source of consciousness (for review and discussion, see Baars, 2002; Cosmelli, Lachaux, & Thompson, 2007; Dehaene & Naccache, 2001; Kouider, 2009; McGovern & Baars, 2007; Palmer, 1999). It is highly reasonable to suggest that indeed the brain's activity is associated with conscious experience and awareness. However, this proposal is based on a correlation between neurophysiological indexes (e.g., neuroimaging) and expressions of consciousness, and is not a mind/body theory. On the basis of an extensive review of neurodynamic approaches to consciousness, Cosmelli, Lachaux, and Thompson (2007) conclude: "The neurodynamical approach works at the level of correlations, albeit refined ones" (p. 763). (Note that it is not easy to apply even the correlation method in research. For critiques of this methodological problem, see Uttal, 2001, 2014.)

A large and principal part of research on conscious experience centers on the question, "How can consciousness with all its properties and kinds be explained?" (see, e.g., Van Gulick, 1995). Chalmers (1996, 1997a) suggests distinguishing two kinds of consciousness problems: easy and hard. He defines these problems thus: "The easy problems of consciousness are those that seem directly susceptible to the standard methods of cognitive science, whereby a phenomenon is explained in terms of computational or neural mechanisms. The hard problems are those that seem to resist those methods" (1997a, p. 9).

The hard problem is expressed in the enormous difficulty in understanding how cerebral, neurophysiological states become consciousness states;

how conscious phenomena, private conscious experiences, are reduced and stem from a neurophysiological system. The easy problem is connected to cognitive behavior, for example, responding to an environmental stimulus, distinguishing, organizing information, reporting on mental states, concentrating attention, and voluntarily controlling behavior. These behaviors are explicable, in Chalmers's opinion, through cognitive and neurophysiological mechanisms, that is, through mechanisms that realize the function of the studied behavior, for example, the function of response to an environmental stimulus, of discrimination, and so on. I agree with Chalmers that the hard problem is indeed hard, but I don't agree that the easy problem is indeed easy. To my mind, it is no less hard than the hard problem, for the following reasons.

A complex behavior (e.g., perceiving, concentrating attention, learning, remembering, planning a project, etc.) is the carrier of a conscious experience—that is, behavior is interwoven with awareness-experience: it involves a sense of reality, an experience of being, conscious existence. Without consciousness, responses, actions, and behavior would be purely motor movements; we would be nothing but robots, machines, moving in the physical world, like the movement of stars, electrical discharges, and the reflection or absorption of light. We would be explainable by a collection of physical, chemical, and biological laws alone; and in truth, there would be no point to living, no motive or reward for what we want, believe, and do. The explanations that Chalmers suggests for the easy problems are nothing but mechanistic explanations, as may be seen from the explanations he offers for access to, and reportability of, mental states: "To explain access and reportability, for example, we need only specify the mechanism by which information about internal states is retrieved and made available for verbal report" (1997a, p. 10).

In his opinion, the appropriate explanation is given by a description of the cognitive, neurophysiological mechanism that performs the function of the studied behavior. These explanations, by their very nature, leave no room for consciousness as an explanatory factor—consciousness is an epiphenomenon. This is the state of affairs that I reject with both hands. (Chalmers's [1997a] article is a target article, which gave rise to a large number of critical papers that resulted in Chalmers's [1997b] response to them [see the book edited by Shear, 1997]. Some of these critical papers, which Chalmers [1997b] terms "non-reductive analyses," assert, like what I have been arguing here, that the easy problems are no less thorny than the hard problems.) According to the MD and MEF approach, a mechanistic explanation alone cannot contend with complex behavior involving consciousness. Without using, in addition to the mechanistic explanation, a mentalistic explanation, the mechanistic explanation will be a partial explanation only—an explanation appropriate for physical–neurophysiological robots. The mechanistic explanation can detail the physical–physiological mechanism responsible for raising the hand, but in no way can it offer an explanation for the most important thing of all, for the questions, "Why was the hand raised? What is the meaning of raising the hand, its purpose? And how does it happen that a motor movement turns into an instrument for expressing a mental state or process? That is, how does it come about that a certain motor movement becomes an instrument in the service of the mentality? While one may suggest that different motor movements can realize a given mental state, the question how this MS has been created has not yet been answered. I suggest, therefore, that meaningful behaviors are a kind of an implementer of the mental meaning. These behaviors should not be seen as a kind of motor behavior to which consciousness has been added, but the reverse: this behavior, in its very essence, is the conscious meaning that has been induced in it.

CAN MENTALISTIC EXPLANATIONS ANCHORED TO THE CONCEPT OF CONSCIOUS EXPERIENCE BE REDUCED TO MECHANISTIC EXPLANATIONS (E.G., TO A NEUROPHYSIOLOGICAL EXPLANATION)?

As I have emphasized in the foregoing chapters, it is hard to establish a parallel between theory reduction (between one theory and another) and reduction among explanatory models, because these are not theories like theories of perception, learning, and remembering, but in essence they are procedures that guide the researchers in how to propose explanations in certain fields. Therefore, the present question focuses more on the possibility of replacing a mentalistic explanation (which as stated is anchored to subjective concepts connected to the individual's inner world, will, belief, intention, feelings and emotions, consciousness, etc.) with a mechanistic explanation (anchored to objective concepts associated with physics, chemistry, physiology, computer processes, and the like). (Note that psychology accepts mechanistic explanations and rejects mentalistic explanations.) First, I shall consider briefly the problems connected to psycho-neural reduction, and then deal with other ways of capturing mentalistic concepts by means of mechanistic concepts, such as substitution of a teleological explanation by a causal explanation. The discussion of all these cases will show that there still has not been found a clear and accepted way of replacing a mentalistic explanation with a mechanistic.

One way to solve the mind/body problem is through use of scientific methodology for inter-theory reduction, that is, through an attempt to reduce psychological theory to neurophysiological theory, something known in the professional literature as psycho-neural reduction. In other words, the question is whether it is possible to explain a psychological theory by means of neurophysiological theory (see, e.g., Bickle, 1998; Kim, 1998; Nagel, 1961; Rakover, 1990; Silberstein, 2002; van Riel & Van Gulick, 2016).

To clarify this question, I shall first describe briefly and schematically the classic methodology for inter-theory reduction, Nagel's (1961) model of reduction. One theory, which we shall call the reduced theory (T_R), is reduced to a second, more basic theory, which is called the basic reducing theory (T_B), when T_R is derived from T_B, together with certain bridging laws, which connect the concepts of these two theories (see also chapter 3). Usually the bridging laws are perceived as identities, for example, in the case of reducing thermodynamics to statistical mechanics, according to which it was proposed that temperature equals the average kinetic energy. In this case, statistical mechanics also offers an ontological (material) explanation for the macro concept of temperature through the micro concept of kinetic energy. Note that Nagel's model of reduction is very close to Hempel's (1965) D-N model of explanation. The main difference is that in the latter model, one deduces with the help of certain initial conditions an empirical phenomenon or generalization, and in the former (Nagel's) model, one deduces with the help of certain bridging laws another theory. Nagel's model aroused several severe and interesting criticisms—for example, the model does not fulfill the requirements for the correction of T_R and for its elimination and replacement under certain conditions (for other criticisms, see, e.g., van Riel & Van Gulick, 2016).

In the case of psycho-neural reduction, identity between a mental state and a neurophysiological state offers an ontological explanation of mental states: these are nothing but neurophysiological states. If the bridging laws are not identical but, for example, correlations between concepts of the two theories, then the reduction will be weak, because these correlations in themselves require explanations and because it is not at all clear if they are nothing other than a kind of accidental generalization. (See different approaches to reduction and discussion on bridging laws in Causey, 1972; Fodor, 1974; Nagel, 1961; Rakover, 1990; Schaffner, 1967, 1993; Sklar, 1967; van Riel & Van Gulick, 2016.)

Several arguments have been put forward against psycho-neural reduction (see discussion in Barendregt & van Rappard, 2004; McCauley & Bechtel, 2001). Most of them stress that psycho-neural bridging laws are not possible. For example, Davidson's (1980) famous argument, called "mental anomalism," is based on the claim that a psycho-neurological law is not possible.

One of the most forceful arguments against psycho-neural reduction is that of multiple realizations (see Fodor, 1974, 1998; and see a different view and debate in Bechtel & Mundale, 1999; Bickle, 1998; Looren de Jong, 2003). To exemplify this argument, we shall discuss the behavioral state of pain.

According to the functionalism approach (which proposes that a mental state can be realized by many different material states [e.g., various neurophysiological states that fulfill the MS's causal role in producing behavior]), pain is a mental property that can be realized by material processes such as various neurophysiological processes found in a large number of organisms (humans, dogs, cats, fish, reptiles, etc.), and also the electrical processes located in highly advanced and elaborate computers. On the assumption that this argument is correct, it becomes impossible to reduce a psychological theory to a neurophysiological theory, because it is not possible to find a bridging law that will join through identity the state of pain and a particular neurophysiological state. If this is the case, a bridging law by means of which it would be possible to reduce psychological theory to neurophysiological theory cannot be found.

As a counterargument, supporters of the reductionist approach proposed a disjunctive bridging law: pain is a disjunctive realization of different material (neurophysiological) states: NS_1 or NS_2 or . . . NS_i. But a disjunctive bridging law, quite simply, is not considered a scientific law. For example, which NS is the most suitable for the realization of pain? Are all material (neurophysiological, etc.) processes in the world suitable for realizing pain? In addition to fulfilling the MS's causal role, are there any additional properties that make a given specific NS appropriate for realization of the MS?

A further argument against psycho-neural reduction is based on the requirement of "equality of units" (see Rakover, 2002b, 2012b). Briefly, it is not possible to discover a psycho-neural bridging law because this law does not fulfill the requirement of "equality of units," namely, identity of the units of measurement on either side of the law's equation (e.g., $MS = f[NS]$): the measurement units of the psychological concepts are entirely different from the neurophysiological measurement units, and no common measuring standard can be found for them that will unite the psychological with the neurophysiological. While the concepts of the neurophysiological theory are measured by means of electrical, chemical, and molecular changes, cognitive or mental theories are expressed in actions measured chiefly by frequency of correct responses and speed of response. So it is hard to suggest how a bridging law may be built to connect the concepts of these two theories. The combination of measuring units of the chemical changes is not equal, for example, to the measuring unit of the response-speed index. One reason is that this index expresses psychological, not physical, time. Also, response speed is an index that expresses actions of a large number of different cognitive processes (connected to perception, remembering, and deciding) that work linearly and in parallel (see discussion in Pachella, 1974).

In the last twenty years, an effort has been made to revive the possibility of performing psycho-neural reduction (see especially Bickle, 1998). One of the fundamental notions of this approach, which Bickle calls the "new wave," attempts to bypass the need to use bridging laws through the application of a methodology of theoretical models, which leads to the construction of a theory analogous (similar) to the reduced theory T_R that is marked T_R*. According to the classic approach, from the basic reducing theory T_B, one derives the reduced theory T_R by means of bridging laws; but by the new-wave approach, on the basis of T_B, a theory similar to the reduced theory, T_R*, may be created without bridging laws. The question, of course, is, how similar are T_R and T_R*? This question led to further theoretical developments, which I shall not deal with here. The new-wave approach gave rise to many criticisms, which queried the idea of psycho-neural reduction (see summary and discussion on these subjects in Barendregt & van Rappard, 2004; McCauley & Bechtel, 2001; van Riel & Van Gulick, 2016). For example, Bontly (2000) finds that even the new wave cannot escape ontological bridging laws, formulated one way or another: "Even the 'new wave' reductionist must first determine whether mental properties can be identified with or built up from neurobiological parts, and as far as I can see, that's just the traditional mind-body problem all over again" (p. 904). Similar points (that the new-wave approach to reduction is not remote from Nagel's model) have been raised by others (for review, see van Riel & Van Gulick, 2016).

Furthermore, McCauley and Bechtel (2001) argue that the new-wave approach does not take into account that scientific development is accomplished by research cooperation between two theories on different levels of description and explanation, for example, cooperation between cognitive theory and neurophysiological theory. They propose a strategy of "hypothetical identities" as hypotheses that connect the concepts of theories of different levels whose chief function is to stimulate and encourage research, for example, between cognitive theory for processing visual information and local neurophysiological processes in the brain (V4 area). Barendregt and van Rappard (2004), who rely on McCauley and Bechtel, among others, stress that inter-theory reduction is not a theory about the mind/body relationship but a methodological position in science that seeks to create a bridge between theories on different levels, and to pave the way to conducting scientific research.

Another attempt to conceive a mentalistic account in terms of a mechanistic account can be illustrated by the following. Consider, for example, a purposive (teleological) explanation: David drove from Haifa to Tel Aviv in order to meet Ruth. This purposive explanation can be equivalently expressed (translated) as the causal explanation: the thought of meeting Ruth in Tel Aviv caused David to drive there. However, the translation is problematic for at least the following two reasons. First, it is generally assumed that similar causes produce similar effects. This assumption does not hold here: while the goal of meeting Ruth can be achieved in many different ways (e.g., David can walk, run, ride (a horse, a donkey), drive (a car, a bike, etc.), fly, and even sail a boat on the Mediterranean Sea to Tel Aviv, a cause (which is the translation of a goal) does not produce such diversity of effects. Secondly, while it is generally assumed that the cause and effect are separate and different events, this assumption is not maintained here. David's will, belief, and action are interconnected. The drive to Tel Aviv is described as a meaningful action that involves David's will and belief, and similarly a meaningful description of each of these three terms engages the other two (for a discussion of this and related issues, see Rakover, 1990, 1997, 2007).

Generally, it can be suggested that although mental processes are involved in behavior, attempts to explain mental causation (how mental properties causally affect motor and physiological behavior) have encountered impassable obstacles. There are well-known problems, which show that mental causation is impossible to account for, such as the problem of "anomalism," which is based on the absence of strict psychophysical laws, etc. (for review, see Robb & Heil, 2014; Yoo, 2015).

IS A COMPLEX COMPUTER ABLE, LIKE US, TO BEHAVE WITH CONSCIOUS EXPERIENCE?

One may safely conclude that research in cognitive modeling or artificial intelligence (AI) has not succeeded in solving the consciousness problem (for reviews, see McDermott, 2007; Sun & Franklin, 2007). The cognitive-modeling (or AI) approach is developed in intimate connection with the functionalism approach (as both are based on computer science). So let us examine functionalism to see whether it does provide a positive answer to the present chapter's main question. As we shall see, it does not.

Functionalism proposes a solution to the mind/body problem. It has been developed, among other things, as an alternative approach to the identity theory (see discussion in Heil, 1998; Lycan, 2003; Rakover, 1990). Since the 1960s, it has been in one way or another the philosophical basis for cognitive psychology. The functionalism approach has many variations. For example, Polger (2004) argues that according to his method of classification, there are over one hundred variations of functionalism, but he discusses six kinds thoroughly (Churchland, 1988, by contrast, deals with four kinds of functionalism). Common to these variations is the attempt to understand a mental state from the viewpoint of its function, the role it plays in the system it belongs to. What distinguishes them is the sort of answer offered in reply to the central question of the given variation. For example, metaphysical functionalism focuses on the question of the nature of the mental state; intentional functionalism focuses on the question of the "about," the "intentionality," of the mental state; explanatory functionalism focuses on the question of the explanatory role of the mental state; and so on.

I shall not enter into a discussion of these matters, but only comment that generally it is possible to characterize functionalism as a theoretical approach according to which a mental state is defined by a functional property, which may be described as connecting the stimulus to the response, as interacting with other states, and as realized materially (e.g., by the brain). Kim (1996) writes:

> Psychological concepts are like concepts of artifacts. For example, the concept of 'engine' is silent on the actual mechanism that realizes it—whether it uses gasoline or electricity or steam. . . . As long as a physical device is capable of performing a certain specified job, namely, that of transforming various forms of energy into mechanical force or motion, it counts as an engine. The concept of engine is specified by a job description of mechanisms that can execute the job. (p. 75)

This notion does not accord with Type Identity theory, which proposes that a given type of mental state is identical to a certain type of cerebral state, mainly because according to the functionalist approach, a mental state can be realized by different types of materials. Functionalism, then, raises one of the most powerful arguments against the possibility of suggesting a psychophysical law according to which the type of mental state is identical with the type of neurophysiological state, just as identity exists between lightning and electrical discharge, or between water and a mixture of gases (H_2O); this argument is called "multiple realizability." To recapitulate, the main idea underlying this argument is that just as the computer program may be run/realized by different types of computers constructed of different materials, so consciousness is likely to be realized by different materials. And just as a certain computing state in the program is characterized by connections to other computing states, to input and output, so a mental state is defined through its relations to other mental states, to stimuli and responses.

So according to the functionalist approach, a mental state is defined through the function that it serves in the cognitive system. And just as the function "to bid farewell to Ruth" may well be realized through a large number of means (saying goodbye, writing a parting letter, waving), so may a certain mental state, pain, be realized through different materials and in different forms (e.g., by the brain of the human, the monkey, the dog, and the cat, and also by a brain made of silicon). That is, a mental state in *X* may be equal to a mental state in *Y*, as long as the arrangement of the (different) realizing materials in both cases is equal in the functionalist respect. The argument can therefore be raised that a realizing material, such as silicon, arranged in a way equal to the neurophysiology of the mental state of fear in humans, will also be the basis of the state of pain. If the argument of multiple realizability is right, the theory of consciousness–brain identity is wrong, because the theory of identity will be hard-pressed to explain, for example, pain in a dog and a cat (see discussions in Bickle, 1998; Fodor, 1974, 1988; Heil, 1998; Kim, 1996; Polger, 2004; Putnam, 1967).

The functionalist approach has several serious problems due largely to the fact that in essence this approach suggests a mechanistic explanation for the behavior. That is, the role of the causal explanation is ascribed not to mental states but to material states, to materials that realize the mental states, for example, to the neurophysiology of the brain. This ascription runs counter to common, everyday sense, which suggests that mental states have content and meaning; that they appeal to a certain subject; that they are located in us in consciousness, in awareness; and that they are to a great extent responsible for our behavior (see criticisms of functionalism in Block, 1978/1980; Polger, 2004).

Here I shall examine one important critique of functionalism and AI: the famous Chinese Room thought experiment (see Searle, 1980). If thoughts are computations, then one can propose that computers can think like us. Searle's thought experiment contradicts this argument (for a discussion of artificial intelligence and other arguments against it, see Hauser, 2016). If we produced a computer of extraordinary perfection, one whose actions were structured identically to the way the actions of a human are structured, by the functionalist approach this superior computer should have consciousness like a human. Is this argument convincing? Searle (1980), and I following him (Rakover, 1999), believe that this argument is incorrect. In his 1980 article, Searle presented his thought experiment, called "The Chinese Room." In essence Searle undertakes to perform all the operations that the computer does when this machine processes the Chinese language. Assume then that Searle, who hasn't the slightest knowledge of Chinese, assumes the function of the computer, enters the Chinese Room—which constitutes an analogy to the computer—obtains input in Chinese characters, processes them according to a book of instructions and guidance in English, does all the computing operations that the computer has done with this input, and produces output in the Chinese language. Specialists in Chinese study this output, and determine that only an intelligent person, who knows Chinese well, would be able to answer all their questions (the input is in Chinese). This last fact shows explicitly that Searle has passed the Turing (1950) test, namely: if we as judges are unable to distinguish a human from a computer by virtue of the answers that these two give to our questions, it will not be justified to attribute to the computer nonconsciousness. That is, if the computer passes the Turing test, we have sufficient justification to attribute mentality to it. (See summary and discussion on the Turing test in Saygin, Cicekli, & Akman, 2000.) This test, then, supports the hypothesis that Searle understands Chinese just as well as an expert in that language. It may be said, therefore, that since Searle acted exactly like the computer, the computer too understands Chinese. But Searle, who is human and endowed with a mind, consciousness, and awareness, like you and me, argues the opposite. He claims that he leaves the Chinese Room with the same level of ignorance of comprehension of Chinese as when he entered it. That is, the hypothesis of functionalism, that a highly sophisticated and perfected computer will develop conscious awareness, has been refuted.

This thought experiment, as may be expected, sparked fierce criticism. I was not convinced by it, and I gave expression to this in a thought experiment that is a variation on Searle's and supports the criticism concerning the functionalist approach resulting from his experiment. In my variation, I started from the viewpoint of a supercomputer that did a simulation of Searle in the Chinese Room. The analysis from this viewpoint showed that the computer is indeed nothing but a machine devoid of consciousness (see Rakover, 1999).

The Chinese Room argument touches on important subjects closely linked to functionalism, and to the Turing test. This thought experiment may be seen as supporting the hypothesis that syntactical processes, which the computer accomplishes, do not generate words and sentences with semantic meaning—otherwise Searle would have emerged from the Chinese Room with a good understanding of Chinese. So the question that arises is this: How is meaning induced into the products of the computations that the brain performs on physical representations? This is a difficult question—very broad and ponderous, which set off many debates that I cannot deal with here, so I shall merely mention relevant allusions.

According to the "symbol grounding problem" coined by Harnad (1990), it is not clear how it was possible that semantic meaning was imparted to a syntactical system of physical symbols, without the process of generation of meaning taking place in a human person's head. In other words, the question is can Searle, or anyone else who does not live in China, does not belong to the Chinese culture, and does not understand Chinese at all, learn Chinese with all the meanings of that language, using a Chinese/Chinese dictionary alone? Several solutions have been proposed for this problem and others connected to it.

In the framework of the functionalist approach and the conception of consciousness as a process of "symbol manipulation" or "information processing" (also called the classic approach), Fodor (1976) put forward the hypothesis of "the language of thought," which posits that mental representations are linguistic expressions created by a basic innate language similar to human language. The most important alternative to the approach to consciousness as a process of "symbol manipulation" is the connectionist approach, the neural network approach. In this framework, computer programs were developed imitating patterns of activation of neural networks in the brain as realizing mental–cognitive processes. Which of these two computational approaches (constituting two different groups of computational models of cognitive systems) is more efficient? This question, too, gave rise to a debate that is still not over. For example, Marcus (2001), who comes from the area of research into language and higher cognitive processes, compared these two approaches and concluded that a relatively straightforward software of the connectionist approach (a simple multilayer neural network [perceptron]) finds it hard to tackle several behavioral problems that classic symbol manipulating manages successfully; that is, the neural network approach does not display adequate power and flexibility. He also suggested that the symbol-manipulating mechanism is innate and developed by evolution. (Interesting discussions on these and other matters can be found, for example, in Block, 1995; Crane, 1995; Fetzer, 2001; Fodor & Pylyshin, 1988; Heil, 1998; Smolensky, 1988.)

Finally, I will attempt to criticize the Functionalists' rationale that the mind–body processes can be viewed as computer (software–hardware) processes. As mentioned above, Functionalism was proposed as a solution to the mind/body problem and formed the philosophical basis of cognitive psychology and the cognitive sciences. According to this infrastructure, the computer, as a machine that performs manipulations on physical symbols, constitutes an extremely important analogy to the mind/brain (or even an identical system to the mind/brain), which enables understanding of psychological mechanisms for processing information such as perception, attention, learning, memory, decision-making, and so on (see Bem & Looren de Jong, 1997; Copeland, 1993; Franklin, 1995; Haberlandt, 1997; Pylyshyn, 1984; Rakover, 1990; Thagard, 1996; Von Ekardt, 1993). Within this approach, I shall discuss the following question: Does the computer analogy help us explain the relationship between consciousness and the brain?

To answer this question, I shall first very briefly review how the computer makes computations, answers questions, prints text, and draws figures (see, e.g., Block, 1995; Deitel & Deitel, 1985; Von Eckardt, 1993). The important point that I would like to start with is that the computer is a mechanism composed of a large number of parts and processes that mediate between the information put into it and the information coming out of it. All these parts and processes are without exception mechanical mechanisms, whose functioning is explained by means of mechanistic models. For our purposes, we shall examine how a program is translated into a series of physical operations, which are performed by the computer according to the laws of electricity and electronics.

First, the software, which the programmer wrote in a higher computer language, automatically undergoes translations from language to language until it is translated entirely into a language based on several simple operations that the computer is able to perform, logical operations done on chains of zeros and ones (0 and 1), namely on binary digits. Let us look, for example, at the how the computer adds up numbers. This addition is carried out by a highly complex combination of three electrical units that realize the following three logical units:

- *And* unit takes two inputs (each input being able to obtain a value of 0 or 1) and produces 1 on condition that the two inputs are 1. As for any other combination—00 01 10—the *And* produces 0.
- *Or* unit takes two inputs and produces 1 where at least one of the two inputs is 1.
- *Not* unit takes one input and changes it from 1 to 0 or from 0 to 1.

Without going into the arrangement of these units into a mechanism of adding up numbers, I shall say only that the mechanism can add any number by translation of the decimal language into binary language, and back into the decimal language that we know well. The point I am stressing in this example is this: the computer is able to perform only these three simple operations. When an instruction is inserted into it to add up two numbers, it translates them into binary language, into series of 0 and 1, inserts these series into the adding mechanism, performs these simple operations (*And, Or, Not*), and in the end yields the desired result.

Secondly, the reason why the computer successfully performs what is written in the software (the above three operations that recurrently appear according to the mechanism for adding up numbers) lies in the connection made between the digits 0 and 1, and between two electrical states in the computer: without electrical voltage (0) and with electrical voltage (1). As a result of this connection, the computer carries out the binary calculation physically (there is electricity; there isn't electricity), namely, what is written in the software.

Hence, higher software language, which is nothing but a procedure that defines certain operations done on physical symbols, is subject to entire and precise breakdown into elementary and simple logical–mathematical routines based on the binary calculation (0 and 1), which the computer performs physically, electrically. So every higher computational procedure (program) is defined wholly by subroutines, which in the end are performed actually and completely by physical operations. By means of these mechanical procedures, it is possible to develop different programs, which, as I mentioned above, are capable of doing calculations of different kinds, of adding up and multiplying numbers, of solving mathematics equations, of word-processing, and of drawing static and dynamic pictures.

Now, let us return to our main question concerning the analogy to the computer: Is the consciousness–brain connection analogous to the software–hardware connection on which the computer rests? A number of researchers and philosophers believe that the answer is affirmative (see, e.g., discussions and critiques in Block, 1995; Cummins, 1983; Dennett, 1979; Rakover, 1990). The idea underpinning this answer is the notion of breaking down an intricate system into its parts: software can be decomposed into simple components (simple routines), which may be disassembled into still simpler parts, until the simplest, the most primitive, units are obtained—such as the units *And*, *Or*, *Not*, binary units that decide between 0 and 1, two states that may be defined by means of two electrical states; so precisely is it possible to decompose every cognitive process, cognitive property, to the most primitive level that may be connected to the neurophysiology of the brain. And just as the computer actually runs the program by means of its electrical operation, so the brain runs the cognitive system by its neurophysiological operation. Dennett (1979), describing this decomposition idea as the "top-down strategy," writes: "A top-down strategy that begins with a more abstract decomposition of the highest levels of psychological organization and hopes to analyze these into more and more detailed smaller systems or processes until finally one arrives at elements familiar to the biologists" (p. 110).

Dennett goes on to describe this procedure of decomposition in an entertaining way. He describes a mental process by means of a flow chart made up of small boxes interconnected by arrows, where every box is viewed as a "homunculus" with certain functions. He continues:

> If we then look closer at the individual boxes we see that the function of each is accomplished by subdividing it via another flow chart into smaller, more stupid homunculi. Eventually this nesting of boxes within boxes lands you with homunculi so stupid (all they have to do is remember whether to say yes or no when asked) that they can be, as one says, "replaced by machine." (p. 124)

Does this description indeed hold? To my mind, the process of decomposition of behavior carrying meaning, conscious behavior, is very difficult, and it is not easy to propose how such behavior may be connected to simple neurophysiological states. Several barriers block the application of the decomposition process.

First, while decomposition is based on a complete mechanical system such as the computer, the consciousness/brain system is not perceived as a purely mechanical system. For example, in the mechanical system, it is possible to point to its basic building blocks, but it is not easy to do so in the consciousness/brain system. Therefore, application of a mechanistic decomposition process to the mind/brain system may run into problems whose solution is not in sight. Fetzer (2001) in his book *Computers and Cognition: Why Minds Are Not Machines* says something very similar: consciousness is not a computer program because for the computer the symbols of the software are nothing but meaningless physical symbols, and because consciousness works on conscious symbols in a manner entirely different from the algorithmic manner in which the computer works on physical signs.

It is hard to understand how the conscious emotional-thought experience "I love my cat" can be broken down into simpler modular parts, to the basic binary computational level, which may be linked to neurophysiological states in the brain, without losing information, without losing the whole conscious meaning expressed in this sentence. For example, even breaking this sentence into separate words will cause every word immediately to lose the meaning of the sentence as a whole. That is, the meaning of each word separately differs from its meaning in the context of the whole sentence. The word *love*, for example, has different meanings in different contexts: love Aviva, love the state of Israel, love eating an egg sandwich, and so on. By contrast, the decomposition process in the computer loses no information, and the translation from language to language down to the language of the machine is complete (otherwise, the computer would not be a computer). It is not possible, then, to break down a meaningful experience without warping it. Therefore, mental behavior is plainly not of the kind of phenomena that may be decomposed similarly into systems such as the flashlight or the computer.

Second, this difficulty is not only characteristic of the above sentence but also characteristic of cognitive psychology as a whole, because the basic concept of this psychology, the concept of information, is wide open and not defined. This concept is defined fully in the computer sciences (by a formula), but in psychology, as stated, it is unlimited, undefined, and applies to everything imaginable in the context of consciousness and of the cognitive system: syllables, words, sentences, content, metaphors, sounds, processes of judgment and deduction, and so on (see Palmer & Kimchi, 1986). This system of psychological information, then, is not a closed mechanistic system, and as a result information at one stage is not preserved in the transfer to processing in the next stage.

Third, even if we accept the identity of neurophysiological state and mental state, we still will not be able to explain the special properties of mental states by means of neurophysiological properties: what Levine (1983) termed an "explanatory gap" will remain. It is hard to comprehend how one may explain the qualitative properties, say, of the conscious experience of seeing the color red by means of physical knowledge of light and neurophysiological knowledge of the visual system (e.g., Jackson, 1982). As mentioned above, Jackson suggests that a scientist who knows everything there is to know about seeing colors but who has spent her entire life in a black-and-white environment will finally learn to understand experientially the meaning of the color red the moment she is exposed to a colorful environment. That is, even if we succeed in breaking down the vision of a red tomato into its components, and reducing them at the end of this decomposition to a number of basic neurophysiological mechanisms, we shall not be able to close the explanatory gap. Therefore, I believe that in this process of decomposition, something important gets lost: the conscious experience of the red color of the tomato.

Fourth, in an article that summarizes and discusses the relevant literature, Looren de Jong (2003) points out several snags in the operation of the decomposition process, for example, the possibility that higher processes possess properties of a totally new phenomenon, "emergent properties," and the components of this phenomenon are very hard to grasp, and that higher processes are continuous and dynamic, so that it is not easy to suggest how they may be decomposed into modular parts.

DISCUSSION

The short review in this chapter has raised serious difficulties that defeat the various attempts to solve the mind–body, consciousness–brain problem, so to this day the mind/body theory has still not been revealed.

I believe that it is hard to find proof showing it is possible or impossible to find a connection between mental state and neurophysiological state similar to a logical or a mathematical proof, because the mind/body problem is firmly connected to the framework of the empirical sciences based on hypotheses and observations. My impression from the above review is that the solutions offered for the mind/body problem encountered the following main obstacles.

Epiphenomenalism

Some of the important relations between MS (conscious experience) and NS developed by modern philosophy focus on the relation of supervenience, which proposes that a change in MS is not possible without a change in NS, and on the relation of realizability, which proposes that MS is realized by NS. These relations create a problem. On the one hand, MS has no explanatory function, since the whole explanatory job lies on the shoulders of NS. (It is very easy to ascribe causal powers to NS.) So the status of MS proves irrelevant, epiphenomenal (i.e., MS is influenced by NS but cannot influence NS or behavior). On the other hand, everyday intuition and common sense tell us that consciousness (MS) has an evident explanatory role in behavior.

Similarly, Kim (1996), discussing these relations with an example from pain, the contraction response, and the neural state (N), writes:

> It is incoherent to think that the pain somehow directly, without an intervening chain of physiological processes, acted on certain muscles, causing them to contract; that would be telekinesis, a strange form of action at distance! . . . It is implausible in the extreme to think that there might be two independent causal paths here, one from N to the wincing and the other from the pain to the wincing. (p.150)

The epiphenomenalism obstacle is apparently an insurmountable hurdle. Kim (2002) writes in his précis for his 1998 book *Mind in a Physical World* the following final conclusion:

> To summarize, then, the problem of mental causation is solvable for cognitive/intentional mental properties. But it is not solvable for the qualitative or phenomenal characters of conscious experience. We are therefore left without an explanation of how qualia can be causally efficacious; perhaps, we must learn to live with qualia epiphenomenalism. (p. 643)

The major argument is that the functional property of an MS does not possess explanatory power in itself, as the explanation is made by a NS, which realizes the function of an MS.

The "Qualitative Gap"

It seems that it is very hard to understand how neurophysiological states and processes become conscious or generate conscious experience. A qualitative gap opens here between the two stances: the everyday knowledge stance (the explanation of behavior has to take into consideration the fact that we are conscious creatures) and the modern science stance (the explanation of behavior must be given within the natural sciences framework).

"Empirical Measurement"

While in science, theories are constructed that describe/explain mechanistically the connection between one NS and another, and between neurophysiological and motor behavior, no theory has been found that is able to describe/explain mechanistically the connection between NSs and MSs or between MSs and behavior. The reason lies, among other things, in the failure to measure conscious phenomena, MSs, as we measure physical and neurophysiological events.

"Causal Explanation"

In a large number of explanations of the mind/body connection, no account is taken of the fact that the causal explanation does not uphold the rule of transitivity (see previous chapters). From the two assumptions (a) is bigger than (b) and (b) is bigger than (c), it necessarily follows that (a) is bigger than (c); but it does not follow from the two assumptions (a) causes (b) and (b) causes (c) that (a) causes (c). For example, from the fact that squeezing the trigger causes the primer to strike the propellant in the cartridge, which explodes and causes the bullet to fly, it does not follow that squeezing the trigger in itself causes the bullet to fly. Owens (1989) suggests that the causal explanation is not transitive because the chain of reasons can come about accidentally, for example, in the case where person A finds himself by chance in a certain place, and there he is murdered by person B. (Owens notes that even Hempel [1965] tends to agree that the causal explanation does not exhibit a relation of transitivity.) Hence from the proposal that (a) neurophysiological processes generate (b) consciousness, and from the proposal that consciousness is involved in generating (c) meaningful behavior, it does not necessarily follow that (a) neurophysiological processes in themselves generate (c) meaningful behavior (these processes can generate only neurophysiological activity or motor movements). Since the causal relation does not always preserve transitivity, one may suggest that there is room for consciousness to play a causal factor in the explanation of behavior (and see Flanagan, 1992; Rakover, 1996; Velmans, 1991).

"Methodological Limit"

What is the "solution limit" that present-day scientific methodology can reach in the attempt to explain the mind/body problem? To my mind, the solution stops at the methodological limit of the correlation between indices expressing the mental and indices expressing the behavioral–neurophysiological. It is possible to construct interesting correlations between neurophysiological activity in the brain and various indices of consciousness and behavior. For example, using functional magnetic resonance imaging (fMRI), it is possible to draw up a correlation between, for example, areas of activation in the brain and a feeling of pain or fear. However, this is nothing more than a mere correlation. Furthermore, even this correlation is problematic: Uttal (2001) in his book *The New Phrenology* sharply criticizes the local correlations in the brain, namely, the hypotheses concerning brain localization of mental processes by use of modern imaging techniques. The reasons for this are, for example, the difficulty of defining and decomposing mental and neurophysiological processes into basic distinct units, and the enormous complexity of the brain/cognitive processes and their information interactions.

Chapter 6

Methodological Dualism (MD) and the Multi-Explanation Framework (MEF) (II)

Setting Up

The present chapter presents my new approach to explaining human behavior (including supreme animals, such as apes, dogs, cats, and dolphins), which takes into consideration the most important and crucial component: conscious experience. This chapter is based on the foregoing chapter 5 (see also Rakover, 1997, 2007, 2011/2012a, 2011/2012b). It consists of the following four sections:

- *First section*—presents the rationale for developing the Methodological Dualism (MD) and the Multi-Explanation Framework (MEF). It is based on a kind of argument, a "justificatory argument," which suggests reasons and justifications for developing Methodological Dualism MD and MEF. On the one hand, this kind of argument is different from the deductive or the inductive; on the other hand, it is also different from a mere justification that is made to suit a given action. The conclusions of this argument are justified by reasoning that handles certain empirical facts.
- *Second section*—presents the major ideas of MD.
- *Third section*—presents the major ideas of MEF.
- *Final section*—presents an overview and a discussion of the major contributions of the present approach and several comparisons with other relevant approaches.

FIRST SECTION: RATIONALE FOR DEVELOPING MD AND MEF

The rationale is based on the following justificatory argument that leads to several conclusions:

Assumptions

1. To date there is no accepted theory that explains the relationship between mind and body, consciousness and brain (i.e., there is no mind/body theory).
2. It is hard (if not impossible) to conceive mentalistic explanations (e.g., based on our own will, belief, intention, consciousness) in terms of mechanistic explanations (e.g., based on events or processes which are physical, neurophysiological, cognitive-computational).
3. In most cases behavior is saturated with conscious experiences. We are aware of our everyday behavior, the environment in which we live, and the relations between us and the environment (physical, social). We are often conscious of our own consciousness, we act to change our consciousness, and we are driven by our state of awareness.

Conclusions

Given these assumptions, if one aspires to suggest a complete account for behavior, the following can be concluded:

1. If one suggests only a mechanistic explanation for behavior (in the manner accepted by Behaviorism, Cognitive Psychology, and Physiological Psychology), the explanation will be incomplete because it disregards consciousness.
2. If one aspires to present a comprehensive explanation of behavior, one has to take into consideration consciousness, that is, to offer mentalistic explanations in addition to mechanistic accounts. This dualistic approach will provide a better account of behavior than the one proposed by a purely mechanistic explanation.
3. Since to date there is no mind/body theory and a mentalistic explanation cannot be founded on a mechanistic one, and since one aspires to offer a comprehensive account of behavior, one has to develop a special model for a mentalistic explanation of behavior in addition to and in coherence with the mechanistic explanation.
4. In view of the above, the MD and the MEF were developed.

As mentioned above, the assumptions of the justificatory argument are supported by many professional publications. I will briefly resummarize the major ideas here (for details, see chapter 5):

- We are still incapable of understanding the mind by means of the brain, and consciousness by means of neurophysiological-computational processes occurring in the brain. To date no one has succeeded in explaining conscious mental states and processes by physical and neurophysiological concepts. That is, we still have no mind/body theory.

- It has been suggested that wide-ranging activity of a neural network that unites various functions in the brain (such as the "global neural workspace") is a source of consciousness. However, this proposal is based on a correlation between neurophysiological indexes (e.g., neuroimaging) and expressions of consciousness, and it is not a mind/body theory.
- Research in cognitive modeling or artificial intelligence has not succeeded in solving the consciousness problem or generating a conscious computer.
- It seems very difficult to conceive mentalistic accounts such as purposive explanation in terms of mechanistic accounts such as causal explanation. One main reason for this is that the term *purpose* in the purposive explanation cannot be translated straightforwardly into the term *cause* in a causal explanation.
- In many behavioral cases, an appropriate explanation of the behavior under study demands that one consider conscious experience as an important component in the *explanans*.

Thus, on the basis of the above, it is safe to accept the foregoing assumptions of the justificatory argument and propose that the stage is set for the development of MD and MEF.

SECOND SECTION: METHODOLOGICAL DUALISM

Methodological dualism does not offer a metaphysical-ontological solution to the mind/body problem. It is a methodological approach that circumvents the mind/body problem by developing means and proposing methods for constructing specific multi-explanation theories of behavior that harmoniously use mechanistic and mentalistic explanatory schemes together. Multi-explanation theories are not just a kind of instrumentalist theories—namely, efficient calculating machines for predicting behavior—but realist theories. The concepts of these theories attempt to represent actual mentalistic, cognitive, and neurophysiological states and processes in the mind/brain system.

Methodological Dualism is based on several cornerstone ideas, which establish the proposal that a mentalistic explanation (a purposive explanation) does meet the scientific methodological requirements. A mentalistic explanation is part of scientific research. Although it cannot be reduced to an acceptable mechanistic explanation, it will be shown that mentalistic explanations are encompassed by the scientific methodology.

Mechanistic versus Mentalistic Explanations

A distinction is made between mechanistic and mentalistic explanation procedures. As mentioned above, mechanistic explanations, which are common in the natural and social sciences, can offer explanations for behavior of humans and animals by an appeal to physical, chemical, physiological, genetic, and evolutionary factors, and also to stimulus–response–consequence relations (without the use of mental terms in explanations), and to cognitive-computational processes analogous to the workings of a computer, such as symbolic (classical) models or neural networks (see, e.g., Bechtel, 2008c; Rakover, 2007). Mentalistic explanations offer everyday accounts for people's behavior by appeal to their mental conscious states and processes (such as will, belief). For example, David drove to Tel Aviv in order to meet Ruth. In this case, the public behavior (David drove to Tel Aviv) is explained by an appeal to David's conscious experience: David's will to meet Ruth and his belief that the trip to Tel Aviv will realize his will. Underlying this and other examples is the assumption that the individual is endowed with mental conscious states and processes (will, belief, purpose, intention, thought, emotion, and the like) that are the basis of a mentalistic explanation.

Mechanistic and Mentalistic Explanations, and Types of Explanatory Theories

These two kinds of explanations have an interesting and close connection to the structure of the theory with which an explanation is provided (on the structure of a scientific theory, see Bird, 2000; Craver, 2002; Rakover, 1990; Rosenberg, 1988; Suppe, 1977a, 1977b). As will be seen below, only one type of theory (type 1) can provide us a complete mechanistic account. Here I shall propose two methodological criteria for distinguishing these explanatory theories: consistency and measurement.

The first criterion, *consistency*, treats the following question: To what extent are the terms in the explanatory theory well defined and not open to different interpretations, and to what extent are the relations between the terms defined unequivocally, so that given a certain stimulus, the theoretical process will yield one response only. For simplicity, I sort theories into two levels of consistency: complete consistency and incomplete consistency.

The second criterion, *measurement*, treats the connection between the theoretical concepts (stimulus, response) that appear in the explanatory theory and the observations (the actual stimulus, response). For simplicity, I sort the theories into two degrees of measurement situations: valid and reliable measurement, and invalid and unreliable measurement. (Briefly, *validity* means that the measurement measures empirically only what is implied from the concept in the theory; *reliability* means that the measuring process measures the same thing at different times. See, e.g., Neale & Liebert, 1986.) (Although previously, in chapters 1 and 5, I discussed briefly the methodological approaches associated with measurement, to paint here a whole picture of my arguments, I recapitulate certain important points.)

In consequence of these distinctions, I propose three types of explanatory theories:

- Type 1: Theories that meet the criteria of consistency and measurement. These are mechanistic theories characteristic of the natural sciences.
- Type 2: Theories that meet the criterion of consistency but not that of measurement. These are mechanistic theories formulated mathematically or by means of computer language, and are characteristic of experimental psychology and of behavioral models.
- Type 3: Theories that do not meet the criteria of consistency and measurement. These are mentalistic theories and are characteristic of what is called "soft psychology" and everyday psychology.

(Note that I do not believe that there are theories that meet the criterion of measurement but not that of consistency. The main reason is this: if the measurement criterion is fulfilled, the consistency can also be fulfilled.)

Given that, my principal argument is this: type 3, mentalistic theories are scientifically flawed because they meet neither the consistency nor the measurement criterion. Since these two criteria are recommended by the methodology accepted in science, one has to take the following precautions when attempting to suggest a mentalistic explanation: when one is asked to explain a given behavior, one first must check very carefully if this behavior cannot be explained by means of a mechanistic theory. Only when the mechanistic theory fails can one appeal to a mentalistic hypothesis for explanatory help. Despite these methodological flaws, a mentalistic theory may succeed in providing us fairly good explanations of the studied behavior. And now I move on to exemplify briefly the above three types of theories.

Type 1

Let us look once again at a very simple explanation that physicists provide for the problem of the free fall of bodies, Galileo's law: $d = 1/2gt^2$, where d is the fall distance, t is time, and g is acceleration of the body as a result of gravity. Given the time of the fall, one can predict exactly the distance of the fall.

Does this law meet these two criteria: consistency and measurement? To answer this question, we must take into consideration the language in which the law is formulated and what this language represents. The language of the law is the language of mathematics, which meets the consistency criterion and which describes a certain functional relationship (a square function) between input (time) and output (distance). In addition, the physical terms and their relations are well defined theoretically in the framework of Newtonian physical theory. Hence the consistency criterion is well fulfilled.

Does this law also satisfy the requirements of the measurement criterion? The answer is affirmative. To support this answer, one has to examine the way physicists connect the concepts in their theory to empirical observations (make a theory–observation connection) by the procedure called "fundamental measurement" (e.g., Campbell, 1953; Coombs, Dawes, & Tversky, 1970; Michell, 1990, 1999).

The theory–observation connection is created by measurement of objects' properties, such as length, weight, and time. Briefly, I shall describe here "fundamental measurement" as a process in which the relation between a certain quantitative property of an object and the unit of measurement of this property is revealed empirically. For example, if we have before us a straight stick of length X and we find that another stick of length M (which we determined as our unit of measurement, e.g., a meter) goes into X from end to end exactly ten times, we have discovered that length X is ten lengths M (that is, $X/M = 10$). Even though at first sight what we have discovered here appears trivial, this procedure has immense significance on which all physics rests. Why? Because the measurement of length upholds all the mathematical properties of numbers; therefore, it arises that what we say by means of numbers will be said also by means of the objects' lengths.

Given the foregoing, one may comprehend immediately that the explanations suggested by type 1 theories are mechanistic and there is no need for assistance by a mentalistic explanation.

Type 2

In several cases theories in psychology are expressed through mathematical languages or languages by whose use one may eliminate inconsistencies. In these cases, one may propose that the consistency criterion is fulfilled. Unfortunately, however, the measurement criterion is not satisfied. To substantiate this, let us examine once again the "Matching law" that handles learning. How can one explain the following (fictitious) experimental finding: a hungry rat placed in a Skinner box (or operant box) with two levers presses lever A three times more than lever B? The answer is based on the application of the matching law to this problem (the law is also named "Herrnstein's law"; see Davison & McCarthy, 1988; Herrnstein, 1961): $B_a/(B_a + B_b) = Rf_a/(Rf_a + Rf_b)$, where B_a (behavior) denotes the frequency of presses on lever A, B_b denotes the frequency of presses on lever B, Rf_a denotes the frequency of reinforcements (food) given for presses on lever A, and Rf_b denotes the frequency of reinforcements (food) given for presses on lever B.

Given that according to the experimental design, the relation between reinforcements is $Rf_a = 1/3\ Rf_b$, it follows from the matching law that the relation between presses is also $B_a = 1/3\ B_b$. And this is the answer to our question above: the rat presses on lever A three times more than on lever B.

Clearly, this law meets the criterion of consistency, as the relations between the conceptual symbols are well defined mathematically. Furthermore, in addition to the wide empirical support that this law has gained, highly interesting and successful efforts have been made to ground the law in conceptual frameworks such as economic theory and signal detection theory (see review and discussion in Davison & McCarthy, 1988). Still, as we shall see immediately, this law does not satisfy the measurement criterion.

As an example, consider the concept of "reinforcement": its measurement depends on the mental, cognitive, and neurophysiological conditions of the participant in the experiment. What constitutes reinforcement in one has no value in another, while in a third the same stimulus is perceived as something disgusting. Contrary to physics, then, an absolute, valid, and reliable definition cannot be suggested for the concept of reinforcement, and it is impossible to point to a unit of measurement of reinforcement with which we can measure the reinforcement power of various reinforcing stimuli such as food, water, heat, sex, and dignity.

In light of this example, I will suggest that in psychology the connection between concepts and observations is not based on the procedure of fundamental measurement as in physics. To support this, we shall look at the way measurement is done in psychology. Psychologists connect concepts and observations by means of an "operational definition"—namely, the meaning of the concept is obtained by detailing the processes of observation and measuring. For example, aggression is defined by the number of words or gestures considered aggressive in a certain culture; fear is defined by the distance of flight from the threatening stimulus or the period of remaining crouched. Intuitively, one can see immediately that there is a gap between the psychological concept and its empirical measurement. Here are two main reasons for the occurrence of this gap.

First, there is the problem of conscious experience: Most of our responses (and those of supreme animals) are not purely motor movements, reflexes produced automatically, but are responses bearing meaning, responses that express the processing of meaningful information that involves consciousness. I raise my arm not as a mere motor movement but in greeting. I walk not because my legs all of a sudden have started to move, but because I want to reach a certain place or because I want to stretch my limbs. In short, our responses, activities, and actions are brim-full of intentions, desires, ambitions, and sense of awareness—that is, conscious experiences.

But a unit of measurement for conscious experience like that which exists for length cannot be found. It is impossible, for example, to take the sense of the taste of lemon (not the physical stimulus, the lemon solution) to define the unit of measurement of the taste of lemon (the "meter" of the lemon, M_L) and by its means to measure the taste of the lemon, L, so that we would obtain, for example, that $L/M_L = 10$, that is, that the taste of L is ten M_L (ten units of lemon taste). It is impossible to take the feeling of love, to define a unit of measurement of love, and say that Jacob loved Rachel by ten and a half units of love more than he loved Leah. Suggestions of this kind sound nonsensical. Just as it is impossible to find a unit of measurement for conscious experience, so is it impossible to find units of measurement for responses that express these conscious experiences. That is, there is no possibility of measuring these responses, as is customary in physics.

What, then, do psychologists measure? They do not measure a response bearing meanings. That is, they measure a motor, neurophysiological, response alone. They measure the raising of the arm, and not its meaning. They measure an action stripped of conscious experience, that is, a bare movement without its most important component: conscious experience.

Second, there is the problem of the requirement of "unit equivalency": While physical theory, the physical law by means of which an explanation is supplied in physics, satisfies the requirement that I call "unit equivalency" (see Rakover, 2002b), psychological theory does not. According to the requirement of unit equivalency, the combination of measurement units on the one side of the law's or the theory's equation has to be identical with the combination of the measurement units on the other side of the law or equation. To clarify this requirement, we shall look once more at Galileo's law: $d = 1/2gt^2$. Since d is measured by the unit of the meter, the expression gt^2 likewise has to be measured by the unit of the meter. Sure enough, a simple algebraic calculation shows that it is: meter = (meter/time^2) × time^2.

By contrast, psychological theory does not satisfy the requirement of unit equivalency. Let us look at the example of the general structure of a theory or a law in psychology:

Behavior = *f*(Stimuli, Neurophysiological processes, Cognitive processes)

Clearly the units of measurement of behavior (e.g., percentage correct responses, speed of response, force of response) are not identical to the combination of units of measurement of the stimuli, the neurophysiology, and the cognitive processes. The number of correct responses is not identical to the physical units of the brain processes (e.g., differences in electric potential), to the units by which the stimulus is measured (e.g., loudness of the noise), or to the units of measurement of cognitive processes (which in most cases are conceived of as information processing). Even though units of measurement for physical, chemical, and physiological processes exist, it is not possible to find units of measurement for the meaning of the response or the information processing (measurement units for conscious experience). Furthermore, the notion of information in psychology is ill defined, and in fact constitutes a kind of a catchword that includes everything imaginable having to do with knowledge—a notion that defies any attempt at definition (see discussion in Palmer & Kimchi, 1986).

Yet here one may raise the following problem. Let us look at the following fictitious and amusing example. Assume that an association has been found between shirt size and intelligence size (IQ): up to a certain age (the teens), as the size of the shirt grows, so does IQ. Let's assume further that the researchers have expressed this connection in a "law," which I shall call "intelligence-shirt": $IQ = kS$, where S signifies the shirt's size, k is a constant, and IQ signifies intelligence. Clearly, this law does not meet the requirement of unit equivalency, because the right side of the equation is measured in centimeters and the left side in number of correct answers in an intelligence test. But if one interprets the coefficient, k, as expressing the intelligence/centimeter relation, one may say that even this law fulfills the present requirement (the equation has intelligence on either side). The problem is that this interpretation is ad hoc—namely, it is a contrived solution tailor-made especially for this particular coefficient, and in no way parallels the interpretations of the coefficients common in the sciences, which are stable, and their estimates do not change in different conditions, that is, they are general.

In light of this discussion, it may be suggested that nonsatisfaction of the measurement criterion projects lack of clarity onto the world of psychological concepts. In that case, what would prevent one suggesting, for example, the following: Einstein's IQ (say 150) equals the IQ of three morons (each with an IQ of 50)? In brief, an awful situation is liable to arise, in which psychological theory will play the mathematics game correctly, but its concepts will not reflect the processes that take place in the individual's cognitive system.

Given this analysis, it is clear that type 2 theories cannot suggest a complete understanding by appeal to mechanistic explanation only and there is a strong need for assistance by a mentalistic explanation.

Type 3

In many psychological theories (including common or everyday explanations) neither criterion (consistency and measurement) is satisfied. As a simple example, consider the following explanation. How can we explain the fact that David traveled in a taxi from Tel Aviv to Jerusalem? The explanation is based on an appeal to David's private behavior: David wanted to meet Ruth in Jerusalem and believed that a taxi ride would realize his wish. It is possible to suggest that this specific explanation is generated from the following scheme or model of explanation (which some researchers call "purposive law" or "teleological law": see Rosenberg, 1988): if *X* has a will, motivation, to achieve a certain goal, and *X* believes that performing a certain behavior will achieve, realize, this goal, *X*'s will, it will be logical for *X* to perform that behavior. (I will designate this scheme [Will/Belief].)

As the foregoing arguments about not satisfying the measurement criterion apply to the present case too (see type 2 above), I move directly to a discussion of the question of consistency. This explanation does not satisfy the requirement of consistency because from the language in which it is formulated no inter-concept connection is obligatory: there is no obligatory connection between the concepts of will, belief, and behavior. The relationship among these concepts is practical, grounded in daily life. It would come as no great surprise if it turned out that although David had the will to reach Jerusalem, and even believed that traveling in a cab would realize his will, he did not put his intention into practice.

As may be seen from this discussion, Galileo's law illustrates for us a perfect mechanistic explanation: it satisfies the two criteria of consistency and measurement. The matching law, too, is an example of a mechanistic explanation, and it seems to meet the criterion of consistency, but it fails, like the other theories in psychology (including cognitive models based on analogy with the computer), the measurement criterion. Despite the great effort to develop behaviorist concepts free of reference to the internal, mental word, I believe this effort has not succeeded. The teleological explanation offers us an example of a mentalistic explanation, one that resorts to the individual's internal world, but that fails on both counts: on the criterion of consistency and on the criterion of measurement.

In this respect, our order of preference for giving an explanation for a given phenomenon must be the following: theories of the first kind that meet the criteria of consistency and measurement; next, theories of the second kind that meet the consistency criterion but not the measurement criterion; finally, theories of the third kind that meet neither the consistency criterion nor the measurement criterion.

Given this analysis of kinds of theories in science, the following question arises: What harm is there in using theories of the third kind? Here is the answer: not meeting the consistency criterion is liable to impair the ability to produce predictions, and not meeting the measurement criterion is liable to impair the ability to connect observation to prediction, that is, the capacity for the empirical test. However, this analysis does not give rise to a proposal that the mentalistic theories have to be disqualified from the outset, but only to a suggestion that these theories should be treated with caution because of the above weaknesses.

In view of the above, one may raise the following question: Why should we use mentalistic explanations? The answer is plain: because there is a very large group of behaviors in humans and animals that cannot be given a complete explanation by an appeal to mechanistic theories. This group of behaviors demands mentalistic explanations (for review, see Rakover, 2007).

Understanding and Mechanistic versus Mentalistic Explanations

In earlier chapters (1 and 4) I suggested that a nonunderstood phenomenon (i.e., the *explanandum*) cannot be accounted for by a nonunderstood explanatory theory (i.e., the *explanans*). In view of the different types of theories above, one may cast doubt on the ability of the mentalistic theories (especially type 3) to provide satisfactory explanations: their nonsatisfaction of the consistency and the measurement criteria may generate obscurity and fuzziness, and as a result the power of the explanation may be greatly reduced. Nevertheless, I believe that this critique carries only a light weight for the following reasons.

First we must ask the following questions: Where does understanding of mechanistic theories come from? The answer is from the theoretical–empirical anchoring. For example, in high school we learn the theories of physics, including the language of mathematics in which they are immersed, the connection between theoretical concepts and observations, the various experiments that support the theory, including the technology that has developed on their basis. That is, understanding of theories that explain physics does not come easy: these are not intuitive theories, but demand specialization lasting years. And only after we understand the physical theories are we capable also of understanding the explanation that these theories provide for the phenomena under investigation.

And now we shall similarly ask the next question: Where does understanding of mentalistic theories come from? The answer is this: a fraction of it comes from the theoretical–empirical anchoring (see, e.g., examples in the preceding chapter and in Rakover, 2007). On the one hand, this anchoring is not strong and stable as in the sciences, precisely in not satisfying the criteria of consistency and measurement. But on the other hand, a major and important part of understanding these explanatory concepts comes from the very fact of these concepts being subjective and conscious. They are understandable to the individual precisely because they are what that person is, his very essence. Let us examine the following example.

We explain David's trip to Tel Aviv by noting that this act of David's (traveling to Tel Aviv) realizes his will to see Verdi's opera *Falstaff*, which is to be performed at the Tel Aviv Opera House. This explanation is based on David's subjective will/belief. Important to stress here are (a) the phenomenon to be explained in the present case is David's action: why David traveled to Tel Aviv; (b) the explaining system appeals to the will to see *Falstaff* and to the belief that a journey to Tel Aviv will realize this will. This will/belief, therefore, is an important part of the explaining system.

Given this, here is the answer to the above question, "Why is this explaining system understood?" In regular and normal everyday life, the individual's will and belief are mental states that are well understood, since to him/her, they are clear and self-evident. These mental states are truly an important part of his/her very being. Nevertheless, one may well wonder if this is really so. May it not be asked what the causes of will and belief are? Psychology of the personality and psychoanalysis show by a very interesting analysis that not always does the individual understand himself, and in certain cases it arises that the individual's will/belief is nothing but an expression of deep and unconscious desires. Is this not the case? My answer is: true, in these cases it is so. What is thought to be understood proves in the end to be not understood. But here I go on to argue the following:

First, one's will and belief are one's own conscious experience; although one may not understand their basic cause, these mental states provide good explanations for one's own and others' behavior.

Second, if one suggests that will/belief is a nonunderstood phenomenon, one has no choice but to appeal to a new explaining system, a system that must be understood (because, as mentioned above, something that is not understood cannot be explained by something else that itself is not understood), by means of which we shall be able to explain the new nonunderstood phenomenon. (Sure enough, Freud built the theory of psychoanalysis as an explaining and well-understood system, one whereby it would be possible to understand mental states that only seemed to us to be understood until Freud revealed that in fact they were not understood and needed a new explanation.) However, here one should take into consideration the following: although (A) *theory x* explains (B) *will/belief conscious states*, and (B) explains (C) *behavior*, it does not necessarily follow that (A) *theory x* explains (C) *behavior*! So in this case one has no alternative but to rely on will/belief as the *explanans* for the studied behavior.

This process by which the seemingly understood turns out to be not understood is a research process characteristic of empirical science as a whole. It emerges that the explaining systems in the natural sciences also undergo changes in understanding; and what once was thought to be understood proves after further research to be a complex phenomenon requiring a new explanation by an appeal to a new understood explaining system. This apparently is an interminable process, constituting an important part of the development of science. For example, Galileo's and Kepler's laws received an explanation in the framework of Newton's theory of gravity, and the concept of gravity received an explanation and a new scientific meaning in the framework of Einstein's relativity theory.

A Mentalistic Scheme (Model) of Explanation

A distinction is made between a specific explanation and an explanation model (procedure, scheme). A model of explanation is a general procedure for creating different specific explanations for different specific empirical observations. Only explanation procedures (models, schemes) that fulfill the methodological requirements of science are approved and accepted by the scientific community. Various models of explanation have been reviewed for the natural sciences (see chapter 2), and several have been discussed for the social sciences (see chapter 3). Here I concentrate on two different models of explanation: the Deductive-Nomological (D-N) model and the purposive (will/belief) scheme. The first, the D-N model, widespread in the natural sciences (see Hempel, 1965, and chapter 2), proposes that a specific explanation (prediction, called the dependent variable in psychology) is deduced from a law (theory) together with particular conditions (e.g., independent variables).

The second procedure (scheme, model) creates specific purposive or will/belief explanations, which are widespread in folk psychology. (For example, David drove to Tel Aviv because he wanted to meet Ruth and believed that a drive there would realize his will.) Here I propose to conceive the will/belief statement "If *X* wants *G* and believes that behavior *B* will realize his/her will, then *X* will perform *B*" as a mentalistic explanation model:

> [Will/Belief]: *If* X *wants* G *and believes that behavior* B *will realize his/her will, then* X *will perform* B. (The [Will/Belief] signifies that the will/belief statement is conceived as an explanation model.)

This is a new proposition, and it is central to the present approach. It immediately raises the following question: Does this explanation procedure satisfy the methodological requirements for explanation accepted in science?

The professional literature can be interpreted as suggesting a negative answer, because it takes the will/belief statement as a social scientific law that can be placed in the D-N model (let us call it the "purposive-law" approach). Some researchers, such as Churchland (1988), Horgan and Woodward (1985), and Rosenberg (1988), phrase a teleological explanation in the form of a purposive law. Rosenberg, who discusses profoundly the teleological explanation as a law in the social sciences, suggests, "This then is the leading explanatory principle folk psychology offers us" (p. 25). Accordingly, the explanation of the observation that David drove to Tel Aviv will be given by an appeal to the D-N model and the (will/belief) purposive law (if *X* desires *G* and believes that *B* will realize *G*, then *X* will do *B*) in the following way:

Assumptions:

1. Law, theory: the purposive law
2. Particular conditions:
   a. David wants to meet Ruth in Tel Aviv.
   b. David believes that a ride in his car will fulfill his wish.

Conclusion/Prediction: David will drive to Tel Aviv in his car.
Explanation: There is a match between the prediction and the observation.

Hence, according to the purposive-law approach, the structure of the mentalistic explanation follows the structure of Hempel's model that I described above. And if this approach is correct, the suggestion that [Will/Belief] is a special procedure for generating specific purposive explanations is incorrect. My counterargument is that the purposive-law approach is incorrect for a variety of reasons, which I shall discuss below. One of my counterarguments is that the will/belief statement (if *X* wants *G* and believes that *B* will realize *G*, then *X* will do *B*) is not a law, an empirical law based on empirical generalization, but is a model, a scheme of a mentalistic explanation. Were it a law, it would be subject to the procedure of empirical refutation. And since to date there have been thousands upon thousands of cases in which specific purposive predictions have been refuted, one may wonder why this so-called purposive law has not been changed or eliminated from the corpus of knowledge of psychology. Why has this refuted and incorrect law not been discarded? Why do people continue to use it? In short, as we shall see below, my proposition is that [Will/Belief] satisfies the accepted methodological requirements of a scientific explanation model and not of a law.

Arguments That [Will/Belief] Satisfies the Requirements for Scientific Explanation

Reasons were developed for conceiving the mentalistic scheme [Will/Belief] as a scientific procedure. It is revealed that [Will/Belief] fulfills the requirements of scientific methodology for providing explanations. Although it is very difficult to reduce a mentalistic explanation to a mechanistic explanation (see chapters 3 and 5), it is shown here that the scientific requirements for explanation are wide enough to encompass the mentalistic explanation scheme [Will/Belief] too. Hence, according to the scientific game rules, mentalistic explanations are methodologically legitimate. In the following I present arguments that support the proposal that [Will/Belief] (a) meets the methodological requirements for a procedure (model) of scientific explanation, and (b) cannot be conceived as similar to a law in the sciences—that is, the will/belief statement cannot be conceived as a law.

(a) Methodological Requirements for a Procedure of Scientific Explanation

Based on the literature on explanation, I propose that an explanation procedure (model, scheme) has four major characteristics (see Hempel, 1965; Lipton, 1992, 2001a, 2001b; Psillos, 2002; Rakover, 1990, 1997; Salmon, 1990; van Fraassen, 1980; Woodward, 2002). I add a fifth requirement called "empirical irrelevance."

1. *General procedure*
   An explanation model (scheme) is a general procedure whereby the researcher proposes specific explanations for specific phenomena. The specific explanation itself, therefore, is perceived as a particular case of the general scheme. This property is maintained in the mentalistic explanation too: the specific explanation that David waves in order to bid farewell to Ruth is a specific instance of a teleological explanation procedure, [Will/Belief], according to which an individual will perform a certain act if s/he believes that this act will realize his/her desire.

2. *Causes and reasons*
   In most cases an explanation model in the natural sciences assumes that the explanation for a phenomenon is associated with a general law, a theory, a mechanism, which proposes causes for the phenomenon's occurrence. Analogously, regarding a human's (or a supreme animal's) behavior, the explanation of behavior is accomplished by an appeal to internal mental processes that give reasons for its occurrence.

3. *Rationality*
   In the natural sciences, an explanation procedure creates from one sort of information (the *explanans*) another sort of information (the *explanandum*) by means of rules of logic (deduction, induction), mathematics, and probability. Thus, the occurrence of the studied phenomenon is expected because it is predicted on the basis of certain information and rational rules. Similar things happen in the case of the teleological, will/belief, explanation. For example, we expect that given David's wish to take leave of Ruth and his belief that waving is the proper response to realize his

wish, it will be only reasonable for David, as a rational person, to wave. However, here the prediction founded on [Will/Belief] is not based on logic, on statistical probability, or on the necessity that derives from a natural law. Rather, it is based on practical reasoning, on the considered opinion of the individual who takes into account, among other things, his ability, the physical, social conditions to which he is subject, and the significance of realizing his will (see Millgram, 2001; Newell, 1982; Samuels, Stich, & Faucher, 2004; Schueler, 2003; von Wright, 1971).

4. *Empiricism*
The specific explanation generated by the explanation procedure must be attached to reality. This enables an empirical test of the theory, the law, the mechanism that the explanation model uses (e.g., the D-N model described above and in the previous chapters). This requirement is also realized for the mentalistic explanation, [Will/Belief]. For example, there is no problem in testing empirically the explanation that David waved his hand as a sign of his wishing to take leave of Ruth. One can test this explanation by presenting certain questions to the hand-waver: Did you wave as a gesture of leave-taking or to ease the pain in your shoulder? Do you know the person you waved to? Please choose from these five pictures the one at whom you waved, and so on.

5. *Empirical irrelevance*
To use the D-N model, one has to set in the model's assumptions various laws and theories (laws of the movement of bodies, laws in electricity or electromagnetism, theories or laws in biology, etc.) and the relevant particular conditions, and derive from them specific predictions. This model, then, possesses the property of being an explanation storehouse for diverse theories and laws from many and varied fields of research (see chapter 2). Similarly, one may suggest that a hypothesis (theory or law) is confirmed or refuted by the familiar Hypothetico-Deductive (H-D) method. This method also possesses the property of being a storehouse for empirical testing of diverse hypotheses, theories, and laws. These properties underlie the present characteristic: empirical irrelevance.

These properties suggest that observations do not relate empirically to the explanation model and the method of testing themselves, but only to the hypothesis, the theory, the law, inserted into these procedures (see also chapter 4). That is, methodologically the empirical observations are not relevant to the explanation procedure (model) and the method of testing; they do not qualify/disqualify the procedures for explanation or testing—that is, they are empirically irrelevant. If observations were relevant to the theory (law, mechanism, hypothesis, etc.), the explanation procedure and the method of testing, then one discordant observation (negative empirical result) would refute, in addition to the theory, the explanation procedure and the method of testing. Therefore, no theory could be tested and used for explanation, since the testing and explanation procedures would be refuted and eliminated. These procedures would be perceived as incorrect, not worthy of scientific usage.

Does the empirical irrelevance attribute also apply to the procedure of explanation: [Will/Belief]? In my opinion, the answer is affirmative. Consider the following example: David wants to meet Ruth in Tel Aviv and believes that a bus ride will realize his wish. Hence, a specific prediction may be proposed that David will travel to Tel Aviv. But David does not travel to Tel Aviv. According to the present attribute, what was refuted was the specific hypothesis about David's travel and not the scientific method of testing or the teleological explanation scheme, [Will/Belief], whereby this specific prediction was generated. The reason for this is similar to what was stated above: as the D-N model and the H-D method continue to generate specific explanations and specific empirical tests, so the teleological explanation scheme continues to produce specific teleological explanations, specific predictions, that deal with other behaviors of David (and of other people). Otherwise, it would not be possible to use any teleologically specific explanation and to put any specific teleological hypothesis to an empirical test. Why? Because in principle one needs only one negative result to refute the specific prediction, and also the method of testing, and the teleological explanation scheme, [Will/Belief], that generated the specific explanation, that is, to view both the H-D method and [Will/Belief] as futile procedures scientifically.

(b) The Will/Belief Statement Cannot Be Conceived as Similar to a Law

The question that arises here is whether the present interpretation of the will/belief statement as an explanatory procedure is the only one. As mentioned above, the answer is no. Several researchers have formulated the will/belief statement in a way similar to a law in the natural sciences—that is, in the framework of folk psychology, the will/belief statement is viewed as functioning as a law (e.g., Churchland, 1988; Horgan & Woodward, 1985; Rosenberg, 1988). Folk psychology refers, among other things, to people's ability to provide explanations for their own behavior and for the behavior of other people by an appeal to mental states and processes. For example, the question, "Why did David travel to Tel Aviv?" is answered with "Because he wanted to meet Ruth." Folk psychology does not just offer explanations for a behavior that has already happened; it also suggests predictions for a behavior that is going to happen. For example, one may predict that David will drive to Tel Aviv to see Ruth.

Philosophers (e.g., Churchland, 1988; Stich, 1983) argue that compared with cognitive psychology or neurophysiology, folk psychology is a spurious science at root that in the end will disappear from the book of science along with all its notions, just as folk theories about ghosts disappeared. Here is what Churchland (1988) writes:

> Folk psychology is not just an incomplete representation of our inner natures; it is outright misrepresentation of our internal states and activities. Consequently, we cannot expect a truly adequate neuro-scientific account of our inner lives to provide theoretical categories that match up nicely with the categories of our common-sense framework. Accordingly, we must expect that the older framework will simply be eliminated, rather than be reduced, by a matured neuroscience. (p. 43)

I don't accept this approach. I reject the view that the concepts of will and belief have roles only in laws or in scientific theories of folk psychology, and suggest instead that these concepts have crucial roles in the framework of the mentalistic schemes of explanation of behavior. For example, the use of the teleological explanation (by means of the concepts of will/belief), which constitutes an important part of the explanation and prediction of behavior in folk psychology, is not a kind of law or scientific theory, but a scheme for explanation by means of which specific teleological explanations for specific behaviors are produced. The question, of course, is what the justification for my claim is. The answer is in two parts: the first considers empirical irrelevance; the second considers the issue of a law in science.

First, empirical irrelevance: If the empirical irrelevance characteristic holds, then [Will/Belief], conceived as an explanation procedure (model, scheme), is not empirically testable, whereas all laws, theories, and hypotheses are empirically testable. Therefore, the will/belief statement may not be conceived of as a kind of scientific law. Churchland (1988) maintains that folk psychology is unchanging because it is fundamentally bad science, and its fate is to disappear from the book of science just as popular theories about ghosts have disappeared. Against this, I argue, in accordance with empirical irrelevance, that [Will/Belief], as an important part of folk theory, is irrefutable, not because folk theory is bad science but because [Will/Belief] is a mentalistic procedure for generating various specific explanations, and as such, it is not affected by empirical results.

Second, law in science: The will/belief statement does not seem to uphold certain criteria of scientific laws (see Swartz, 1985; Weinert, 1995; Woodward, 2000, 2003). In addition to these, I offer the criterion of "unit equivalency" (see Rakover, 2002b) as another reason why is it difficult to conceive the will/belief statement as a law (let's call it purposive law or "will/belief law" for short).

I shall first examine whether will/belief law upholds the generally accepted and important criteria that distinguish scientific laws from accidental empirical generalizations. To substantiate this distinction, I shall draw a parallel between a known scientific law, Newton's law of gravity, and an accidental empirical generalization, which I shall call "the law of Ruth's party." The Ruth law is based on the observation that everyone who was at Ruth's party last Friday had an IQ higher than 130. Then I shall show specifically that will/belief law, as a law in folk psychology, does not uphold several additional properties characteristic of a scientific law (see Rakover, 1990, 1997).

a. *Counterfactual situations*: What will happen if we throw a stone up into the air? What will happen if we discover a new planet? In these cases, the scientific law supports a fact that has not occurred: the stone will certainly fall to the ground, and the new planet will behave according the Kepler's laws, which can be derived from Newtonian theory. However, it is clear that the IQ of at least one of the participants at future parties given by Ruth is liable to be below 130. That is, the Ruth law does not support a fact that has not happened, being counterfactual (that everyone who will be at Ruth's future party will have an IQ higher than 130).

   It is reasonable to suggest that will/belief law will not support a counterfactual possibility either (that X will realize his/her wish, W_1), because, for example, X will find out that he/she does not know how to perform the action that will realize his/her desire W_1. Rosenberg (1988) discusses these things and finds that will/belief law without the addition of *ceteris paribus* (the condition of all other things being equal) is false. In other words, to make will/belief law efficient, we must add to it a long list of factors that have to be held constant, for example, that compared with other wishes W_i, X's (W_1) is the dominant. The problem is that this list of additions is very long, far in excess of what is acceptable in the natural sciences. For example, Galileo's law requires the falling of bodies in a vacuum. (Interestingly, the formulation of will/belief law in Churchland [1988] and in Horgan and Woodward [1985] already included several factors that are held constant.)

b. *Explanatory power*: A scientific law has the power to explain empirical phenomena. Why did the stone fall to the ground? Answer: because of the force of gravity. This does not hold with the Ruth law. On the assumption that the entry threshold to the law faculty is an IQ of 130, would anyone in their right mind admit to this faculty Michael, for example, simply because he attended Ruth's Friday party? The answer is obvious. Here we should add that the explanatory power of a scientific law stems from its fitting into the broad theoretical–empirical framework. (For example, the law of gravity is connected to Newtonian theory, the laws of Kepler and Galileo to Copernicus, and to an impressive collection of observations and results of experiments.) What is the theoretical–empirical basis of the Ruth law? Clearly, this law is not grounded in a theoretical–empirical network, like the law of gravity.

Similarly, will/belief law does not have the same necessary explanatory power as the law of gravity: the behavior moves from one's wish into practice by means of practical considerations and reasons. The theoretical–empirical connections, into which will/belief law fits, are not the kind of firm connections characteristic of the laws of the natural sciences.

c. *Universality*: A scientific law must be generalizable beyond time and space. We assume that the law of gravity acted on the solar system and other systems a billion years ago, and will act on these system a billion years hence. Nevertheless, all the laws of science are bounded by certain limits, by a certain physical system, and constitute an expression of abstraction and physical idealization. For example, the Newtonian law of gravity is restricted to terrestrial speeds and to objects that are not of atomic or subatomic size; and the laws of biology are limited to various evolutionary groups—for example, explanations of the amoeba's behavior cannot explain cats' behavior, and certainly not the highly complicated behavior of humans. The law of gravity refers to Earth ideally as a point of mass, without taking into account that Earth is constructed of different layers of mass that are not distributed symmetrically. Yet it is fairly clear that while the law of gravity applies to all the stars in the Milky Way, it would be hard to maintain that the Ruth law applies to other parties (of Ronit, Sue, and David) and to other people at Ruth's parties.

This conclusion also pertains to Woodward's (2000, 2003) suggestion that in the special sciences (e.g., economics, psychology) there is no point in talking about laws in the usual sense in physics, but about stable empirical generalizations that do not change beyond certain relevant variables. For example, the matching law that I discussed above may be taken as a stable empirical generalization that applies to different subjects, different reinforcements, and different tasks (experimental designs). Compared with the matching law, it is impossible to perceive the Ruth law as a stable empirical generalization because at the next party it will already be apparent that at least one of the partygoers has an IQ below 130—Ruth's new boyfriend. It is reasonable to suggest that will/belief law does not meet the requirement of universality or stable empirical generalization, because this law, without the proviso ceteris paribus, is false, and it changes extremely over a large number of relevant variables.

Given these, it seems safe to conclude that will/belief law does not satisfy the criteria for a scientific law. Here I shall continue and show specifically that will/belief law possesses certain properties that do not characterize a law.

- The concepts of will (desire, motivation), belief, and action show a certain interdependency, certain logical connections. For example, David's behavior is not just a motor movement but a behavior, an action carrying meaning—David's desires and beliefs. By contrast, in the law of free fall of bodies, the distance of the fall is a term measured methodologically independently of the term time, the duration of the fall.
- The structure of will/belief law is different from the structure of a law in the natural sciences. The concepts that appear in Galileo's law are one-dimensional (time, distance); other concepts in physics are combinations of one-dimensional properties (acceleration, work, energy). By contrast, the three concepts that appear in will/belief law are names of very complex behavioral categories: there are different kinds and varieties of wishes, beliefs, and actions. It is not clear how one may decompose these concepts into one-dimensional components.
- Measurement units for the concepts of will/belief law—will, belief, and action—do not exist, as they do for laws in the natural sciences. Moreover, will/belief law does not satisfy the requirement of "equality of units." As I showed above, in the function Action = *f*(Will, Belief) the combination of the actual indexes of will and belief is not identical to the combination of the indexes of action.
- Given the will/belief law, it is not possible to derive logically, mathematically, the conclusion that *X* will perform *B* from *X* wishes to achieve *G* and believes that *B* will realize his/her goal. The reason is that the transition from the assumptions to the conclusion in the case of will/belief law is not done logically, mathematically, as is the case in the Hempelian explanation, but this transition depends on practical reasoning.

Given these arguments, I conclude that it is hard to conceive will/belief law as a scientific law. It is best to view it as an explanation procedure, [Will/Belief], which generates different specific purposive accounts. Similarly, Von Wright (1971) suggests:

> Practical reasoning is of great importance to the explanation and understanding of action. It is a tenet of the present work that the practical syllogism provides the sciences of man with something long missing from their methodology: *an explanation model in its own right* which is a definite alternative to the subsumption-theoretic covering law. Broadly speaking, what the subsumption-theoretic model is to causal explanation in the natural sciences, the practical

> syllogism is to teleological explanation and explanation in history and social sciences. (p. 27, my italics)

In this passage I have emphasized the words "an explanation model in its own right" because here von Wright expresses an idea similar to the fundamental thesis of the present chapter: will/belief law is not a law, an empirical generalization, but a procedure, a scheme, a model, for producing specific teleological explanations.

THIRD SECTION: MULTI-EXPLANATION FRAMEWORK

In the foregoing section on Methodological Dualism the ground was prepared to develop the MEF. First, it was clarified that good understanding of behavior requires the use of two kinds of explanatory models: mechanistic and mentalistic. Secondly, it arises that [Will/Belief] as a mentalistic explanatory model meets the requirements of scientific methodology. This model may, therefore, be seen as a procedure for creating specific teleological explanations for behavioral phenomena firmly interwoven into the inner world of the individual. The present section, then, is a natural continuation of what has been achieved with the development of the MD. The question before us here is this: How should a specific theory be constructed to explain a given behavior that will be based on the use of two kinds of explanation: mechanistic and mentalistic? The development of such a specific theory depends on the preparation of a theoretical framework, the MEF, which is based on the following ideas.

Explanation and Empirical Testing

The MEF is not a specific theory in a specific psychological area, but is a framework that allows one to develop a specific theory in a specific area of psychology. It suggests guidelines and procedures how to construct a specific and coherent theory, which (a) is based on two kinds of explanation (mechanistic or mentalistic), and (b) can be tested empirically. Although most current psychological theories are based on mechanistic explanations, nearly all behaviors need to be accounted for by appeal to both mechanistic and mentalistic explanations.

Explanation: There is a great difference between the way an explanation is offered in the sciences and the way an explanation should be offered according to the present MEF approach. To clarify this point, we shall study the two following examples, and answer the basic question of how to present a mechanistic explanation for the action of an electric kettle and the operation of a flashlight. Then we shall clarify if it is possible to apply this methodology of giving an explanation to behavior.

I distinguish simple phenomena, such as the free fall of bodies, which may be explained by an appeal to one natural law, from complex phenomena, such as the eruption of a volcano and how a car, an airplane, or even an electric kettle or a flashlight works; the latter are explained by an appeal to several natural laws or theories. For simplicity, we shall look at the familiar working of an electric kettle and a flashlight. How may we explain the boiling of the water? To do so, we must disassemble the working of the kettle into three main subsystems connected with electricity, heat, and water. By means of the laws of electricity, we explain how electricity that passes through a resistance creates heat, and by means of the appropriate laws of thermodynamics and chemistry, we explain how the heat rises from the bottom of the kettle and boils the water. Hence, the kettle's action is accounted for by an explanation of the action of each of its components and by the proper combination of the action of these components, namely, energy transformations (e.g., electricity to heat), a combination that in the end leads to the behavior that we want to understand: the boiling of the water. This explanation, evidently, is based on a chain of different processes (acting one after the other and simultaneously), all of which are explained by mechanistic explanatory model(s) that for its/their purposes use(s) various natural laws relevant to electricity, heat, and water.

Fundamentally, the explanation of the flashlight is similar to that of the electric kettle. How does the former instrument work? To answer the question, we perform the following theoretical–empirical analysis. We dismantle the flashlight into its parts: switch, battery, electric lead, bulb. We explain each and every part and their interactions: when the switch is operated, the electric circuit is closed and the current passes through the filament in the bulb (e.g., a tungsten wire), heats it, and as a result the bulb emits light. Here, too, one observes energy transformations, which are accounted for by mechanistic models of explanation: the action of the battery is based on chemical and physical processes; the action of the filament is based on its functioning as an electrical resistance. We explain the connections across the different parts as processes of conversion of energy: chemical energy converts into electrical energy, which converts into heat, which emits light. This energy conversion can be calculated precisely and explained by means of the appropriate laws, theories, which are set in the appropriate mechanistic models.

As can be understood from these two examples, the main difference in the explanation of simple and complex phenomena is this: in the latter case, one needs to know (a) how many mechanistic theories (laws) to use, and (b) how these explanatory theories interact. In the case of simple phenomena (e.g., free fall of bodies), we set in the explanatory model (e.g., the D-N model) one natural law appropriate for the simple phenomenon under study; in the case of complex phenomena (e.g., electric kettle, flashlight), we set in the mechanistic explanatory model(s) different natural laws, theories, mechanisms, which match the components of the complex phenomenon, namely, which are appropriate for the explanations of the components of the studied phenomenon. The explanatory connections across these components (and their mechanistic explanations) are made through other mechanistic explanation(s) that express transformation of energies among the components so as to unify them as one functioning system.

In view of the above, one may suggest the following relationship among models of explanation, theories, and the complex phenomena in the sciences:

Explanation—sciences: One or several explanation models encompass various theories. These theories are connected or interact in a certain order, so as to generate a satisfactory unified explanation of the studied phenomenon.

In a way similar to the above, I distinguish simple behavior, namely, that which can be explained by an appeal to mechanistic explanation(s), from complex behavior (humans and supreme animals), which has to be explained mechanistically and mentalistically. Can the above methodological approach (which has been developed for the sciences) be applied to the explanation of complex behavior? My answer is no.

The chief reason for this negative answer is that we do not have a theory of transformation from the mind to the body and the reverse—we do not possess mind/body theory. So we are unable to explain how mechanistic components connect to mentalistic components, and the reverse. And therefore we are unable to propose for complex behavior a unified explanation, as in the case of the electric kettle and the flashlight. For this reason, we must turn things inside–out: while in the sciences one inserts into the same explanatory model(s) different theories to explain different phenomena, in psychology we place in the same theory different explanatory models—mechanistic and mentalistic.

That is to say, in contrast to the explanations in the sciences, according to MEF, behavior has to be understood through an appeal to a theory that employs coherently several explanation procedures—mechanistic and mentalistic.

For example, a complete explanation of David's farewell behavior is different from the unified explanation given to a flashlight. As mentioned above, the explanation for the latter is provided by decomposing the flashlight into parts (battery, bulb, etc.), by providing an explanation for the operation of each part through the appropriate mechanistic theory, and by giving an explanation for the interaction of these parts through calculation of transformation of energy. This kind of explanation cannot work with an MEF explanation of David's purposive behavior—the farewell behavior—because we still do not know how mental processes (e.g., David's will, belief) interact with neurophysiological processes (e.g., David waved his hand), how consciousness emerges from the brain. In short, we still do not possess a mind/body theory; therefore, we cannot provide a complete and unified explanation of behavior as we can with the flashlight. One has to suggest an explanatory theory for David's behavior that includes a mentalistic explanation of David's inner world, and a mechanistic explanation for his public behavior.

In view of the above, one may suggest the following relationship between models of explanation, theories, and the complex phenomena in psychology:

Explanation—psychology: A purposive theory employs several explanation models (mechanistic, mentalistic), which cannot be connected, since as yet no mind/body theory exists. As a result, although the explanation given by this theory is improved, a complete unified account for the studied behavioral phenomenon cannot be provided.

Note that this situation in psychology is probably an important reason why the field is fractionated and has not succeeded in developing in any area (e.g., perception, learning, cognition) a top theory like Newtonian theory, the relativity theory, or quantum theory in physics (see Rakover, 2012b).

Empirical testing: While the natural sciences and the present approach differ methodologically in providing explanation, they do not differ in the methodology of employing an empirical test of a theory. No problem hinders testing each theory that explains the functioning of each component in the electric kettle or the flashlight: the intensity of the heat produced by the electrical resistance in both the kettle and the flashlight; the time taken for the water to boil; the intensity of the light produced by the filament in the bulb; and so on. In each case one only has to do the appropriate calculations and compare the prediction with the appropriate empirical measurements. Similarly, no problem prevents testing a specific theory based on MEF (an MEF theory): it can be tested in a way similar to testing a theory in the natural sciences. The main reasons for this are as follows.

The testing method (the H-D method) is indifferent to the kind of theory the researcher uses to explain the experimental results. The H-D method can be used as long as a prediction can be derived from the theory under study (by employing certain rational means) and the prediction can be compared with the observation. For example, there is no problem testing empirically the will/belief explanation that David waved his hand as a sign of his wishing to take leave of Ruth. Since David is acquainted with Ruth, one can predict that he will recognize (choose correctly) her photo out of ten different photos; and Ruth, who will recognize David's photo as well, will confirm that she saw David waving his hand to say goodbye, and also that she waved back, and so on.

The Explanation Suggested by the MEF

As can be seen from the above, one cannot provide a complete explanation for behavior. The reason, as mentioned above, is that as yet no mind/body theory exists, akin to the mechanistic theories of energy transformation so that consciousness will be converted to neurophysiological processes, and the reverse. This, however, does not mean that we should be content with a purely mechanistic explanation only. Why? Because our understanding of complex behavior will be much better when the mechanistic explanation is accompanied coherently and supported by a needed mentalistic explanation. The question is, how should one construct an MEF theory that will be methodologically permissible and satisfactory? The solution suggested here is provided by the development of a suitable method of creating a match between the type of explanation (mechanistic, mentalistic) and the type of behavior under study.

As an example of creating a match between an explanatory model (mechanistic, mentalistic) and behavior, we may analyze the following case of "David goes to the cinema": David wanted to go at seven in the evening to see a particular movie at the Tadmor cinema, and sure enough, at 6:00 p.m. he left his apartment and drove to the movie theater. By the present approach, this case can be seen as based on teleological behavior, so the explanation for David's general behavior will follow the [Will/Belief] path: because David *wanted* to see a certain film at 7:00 p.m. at the Tadmor cinema, and *believed* that a drive to the Tadmor cinema would realize his will, he drove to the Tadmor cinema at 6:00 p.m. This explanation is quite good for anyone who is content with the answer to the question of *why* David drove at 6:00 p.m. to the Tadmor cinema. However, as may at once be seen, this is not a satisfactory answer to the question of *how* David realized his will. Clearly, a routine action of this kind requires action and synchronization among several systems (and subsystems) activated in David. Let us look at several salient systems that David required in order to accomplish his purpose: a system of memory that at six he had to leave the house; motor and sensomotor systems linked to actions such as walking, locking up his apartment, taking out the keys to start the car, driving to the parking lot of the Tadmor cinema, parking, purchasing tickets, entering the cinema, finding the seat, sitting down, and watching the movie—all these actions also necessitate retrieval of items of memory not only connected to finding the correct route to the cinema, buying tickets, handing over the correct sum, and finding the right seat in the cinema hall, but also connected with the performance of automatic actions (which are not located in David's consciousness) such as walking, driving, buying tickets (including remembering/understanding the monetary value of currency notes), and so forth. But for us, the present description will suffice, and we shall not add further details to it. This description too gives a picture of rather elaborate and complex activity. Are all these systems subject to explanations by means of the [Will/Belief] explanatory model? Of course, the answer is no. For example, retrieving the correct information from memory at six o'clock, at the moment of buying the tickets (the sum of money that the coins and the notes represent), and of course the monitoring of all the motor actions that David performed—all these are behavioral phenomena that are very difficult to explain by means of [Will/Belief]. To retrieve information from memory, for example, the proper mechanistic explanatory models have to be sought (e.g., a cognitive–neurophysiological model of memory based on the operation of the computer). Another part of David's activity is connected to a subsystem explained by means of [Will/Belief]. For example, the drive to the Tadmor cinema requires a spatial sense of direction that may be explained as follows: David *wanted* to reach the Tadmor cinema and *believed* that driving along a road known as Cinemas Road would take him to the cinema complex in which the Tadmor was; so David drove along Cinemas Road and arrived at the Tadmor cinema parking lot. I believe that this analysis is sufficient to reach the following conclusion: a complex behavior such as David's is based on chains of behaviors, some of which are explicable by an appeal to mechanistic models of behavior and some by an appeal to mentalistic models. Although a good explanation for David's general behavior by means of [Will/Belief] can be proposed, the complete explanation suggests displaying the entire mechanism responsible for realizing David's will, that is, itemizing the components of the behavior and the match of these components to explanatory models (mechanistic, mentalistic). It should be stressed that by this approach it is indeed possible to explain each and every component in the individual's response phenomenon, but it is not possible to explain how these components, mechanistic and mentalistic, join together because a mind/body theory has still not been found. What I can suggest instead, as a basis for the MEF theory, is an answer to these two questions: (1) How is it possible to match an explanatory model to a behavior and its components? and (2) What is the relation between kinds of explanation for behavior overall and kinds of explanations for the components of this behavior?

Explanation/Behavior Match

The MEF theory is based on a match between explanation procedures (mechanistic or mentalistic) and behavior (we may call it "explanation/behavior match," for short). While there are behaviors that can be accounted for satisfactorily by appeal to mechanistic explanations only (see also chapters 2 and 3), many behaviors and their components need to be approached by both mechanistic and mentalistic explanations. The coherence of a specific theory based on the MEF is achieved by matching appropriately the explanation procedure to the behavior and its components by means of several guidelines and procedures, to be detailed below.

To accomplish this matching, one has to break a behavior down into its components and fit the appropriate mechanistic or mentalistic explanations to the whole behavior and its components. This involves several steps. (1) One determines whether the studied behavior matches mechanistic or mentalistic explanation procedures. (2) One chooses from several explanation schemes the particular one that fits the studied behavior and its components. I call the matching explanation/behavior pair an "explanation unit." (3) Finally, one organizes coherently the explanation units in the specific theory developed for the studied behavior.

How, then, does one perform these steps and fit the explanation procedure (mechanistic or mentalistic) to behavior and its components? Because the answer depends on many factors (associated with theoretical and empirical knowledge), a general solution (a kind of a formula) to this issue cannot be proposed. Accordingly, I shall describe in brief two important markers suggesting when the explanation of a given behavior will be achieved with the aid of a mechanistic explanation procedure and when with the aid of a mentalistic procedure.

(I) *Empirical markers*: Numerous studies discuss the following questions. Is one capable of being aware of the five events connected to one's own behavior: the presented information (the stimuli: verbal, visual, and the like), the response to this information, the mechanism that produces the response, the purpose of the response, and the result caused by the response? Can one control (initiate, change, stop) these five events? The answers to these questions may guide the researcher in the choice of the proper explanation (mechanistic, mentalistic). Here are several examples illustrating the above five events. They start with behavior that is explained mechanistically and end with behavior that has to be accounted for mechanistically and mentalistically.

Note: In many cases the individual is not aware of, nor controls, the five foregoing events, or most of them. These events are associated with chemical, neurophysiological processes in our brain and our body and are accounted for mechanistically (see Morsella, 2005; Palmer, 1999; Rakover, 1983). Many empirical findings are accounted for by appeal to neurophysiological processes. For example, Johnson (1972) shows that several aggressive responses are controlled directly by an electrical stimulation in the cat's brain (in the hypothalamus) and also in the brain of other animals (e.g., monkeys); and LeDoux (1996) shows that an important part of fear behavior (immobility, hair bristling, change in blood pressure) appears without awareness and is controlled directly by a brain area called the amygdale. These kinds of behavior call for mechanistic explanations.

Note: A number of cases are characterized by cognitive impenetrability (Pylyshyn, 1984): if a given behavior is not affected by a change in the individual's goals, beliefs, desires, thoughts, and knowledge, it is reasonable to assume that it is based on innate mechanistic processes (see also Fodor, 1983, on modularity and encapsulation). As an example, let us look once more at Müller-Lyer's illusion from Figure 5.1. Although we know that the right-hand line is equal in length to the left-hand line, our perception says clearly that the right-hand line is shorter than the left-hand one. That is, this perceptional phenomenon is not influenced by our knowledge. This calls for a mechanistic-computational kind of explanation.

Note: In many cases one plans a behavior that will satisfy one's conscious will/belief, as in the case of "David goes to the movies." In fact, many of one's actions (such as reading, listening to music, watching plays and movies, travel and touring, meeting friends, and the like) are performed to change one's own conscious experience. Furthermore, a change in one's conscious-experience is the target of a large part of social interaction. For example, a lover acts in such a way that the courted woman will change her attitude and begin to love him as much as he loves her.

(II) *Marker by "the principle of explanation-matching"*: As mentioned above, this principle deals with cases where a whole behavior is divided into its components, each of which has to be assigned the appropriate (mechanistic or mentalistic) explanation. The following question then arises: What is the relation between the kind of explanation that has been matched to a whole behavior (*A*) and the kinds of explanation matched to its behavioral components (a_1, a_2, a_3, etc.)? To resolve this problem, the principle of explanation-matching is proposed:

- *Mentalistic behavior*: Behavioral components of a mentalistic behavior (a whole behavior explained mentalistically) are likely to receive both mentalistic and mechanistic explanations.
- *Mechanistic behavior*: Behavioral components of mechanistic behavior (a whole behavior explained mechanistically) will receive only mechanistic explanations.

Accordingly, this principle posits that the components of a mechanistic behavior cannot receive mentalistic explanations, while the components of a mentalistic behavior may receive mentalistic as well as mechanistic explanations. Hence, if mechanistic behavior A is broken down into two behavioral components a_1, a_2, where a mechanistic explanation is matched to a_1 and a mentalistic explanation is matched to a_2, then either behavior A was not purely mechanistic, or the match of the explanation to the behavioral component a_2 failed.

For example, since the operation of a flashlight is accounted for by a mechanistic explanation (electricity), the operation of its components (such as battery and bulb) can be given only a mechanistic explanation. To propose a mentalistic explanation is sheer nonsense. As another example, consider the following: since the very popular Müller-Lyer illusion (see above) is not affected by mentalistic factors such as knowledge of the structure of the illusion, and since it appears in fishes and chicks too (see, e.g., Coren & Girgus, 1978), the illusion's explanation has to be mechanistic. So it is hard to see how the components of this illusion are likely to receive a mentalistic explanation. However, if a certain plan of action is put into practice (e.g., David goes to the movies—described above), a complex network of systems (mentalistic and mechanistic) is activated that enables going to the movies and all that it involves. In this case the whole behavior is explained mentalistically, and its components are explained mechanistically and mentalistically.

Problems Generated by an MEF Theory and a Suggestion for Their Solution

The idea that an MEF theory (i.e., a specific theory based on MEF) rests on several explanation schemes (various mechanistic and mentalistic schemes) is liable to give rise to the following three problems, for which a solution is offered by the procedure of fitting the appropriate explanation scheme to the behavior, that is, by the explanation/behavior match.

1. *The ad hoc explanation problem:* Because an MEF theory is based on a number of explanation models, the following problem can be raised. If a certain behavioral phenomenon is not explainable by scheme A, one may try scheme B, and so on, until a satisfactory explanation is found. The problem created by this possibility is that the theory may propose ad hoc explanations and not be empirically refutable. The reason for this is that an appropriate (and a legitimate) explanation can be fitted to any new behavioral phenomenon, and as a result the theory will be protected from empirical testing.
2. *The inconsistency problem:* Since an MEF theory is based on a number of explanation schemes, it may provide an explanation for a certain phenomenon through explanation model A, which may run counter to explanation model B. That is, while explanation A provides a prediction that matches the studied phenomenon, explanation B predicts the opposite. Thus, a theory based on this approach may be beset by internal contradictions.
3. *The incomparability problem:* Two specific theories based on the MEF may not be comparable, as they may employ different explanation schemes (procedures). As a result, it may be difficult to decide whether

dissimilar predictions are generated by the different content of the theories themselves or by the different explanatory schemes they employed. For example, suppose that a certain theory T_1 predicts P_1, whereas T_2 predicts P_2, and T_2 is refuted. The problem is this: while T_1 employs explanation model (I), T_2 employs explanation model (II). Given this, one may ask: which of the two, T_2 or explanation model (II), is responsible for the falsification? Note that the present problem may be seen as an extension of the so-called Duhem's problem; accordingly, it cannot be known what is to blame—the theory or auxiliary hypotheses, since the prediction, which is refuted by the observation, always derives from a theory and auxiliary hypotheses. (See Duhem, 1966, and a proposal for a practical solution in Rakover, 2003.)

It should be noted here that these problems do not arise in physics, because there, for a given research area, one uses the same explanation model (and, of course, one empirical testing method) to treat many physical theories (hypotheses, etc.). To substantiate this, let us examine Galileo's law once again: $d^G = 1/2gt^2$. Now let us assume, to illustrate my point, that in Galileo's day an amateur researcher, Pinocchio, proposed a different law for the falling of bodies, Pinocchio's law: $d^P = 1/2gt^3$. Using the D-N model, this law explains very well the fall of the body in the first second, as does d^G. But what happens after two seconds? According to Galileo's law, d^G, the body will fall $d = 19.6$m, and according to the new law, $d^P = 39.2$m. Obviously, using the H-D method, Pinocchio's law is refuted and Galileo's law obtains empirical support; the new law is not able to explain the results, while Galileo's law does explain them. The point to be emphasized here is this: one is allowed to reach the latter conclusion (Galileo's law does explain the results, and Pinocchio's law doesn't) because in both cases the same explanation model and testing method were employed. So the differences in the explanations (predictions) are caused by the difference in the theories and not by the model of explanation, which was held constant.

The main idea for solving these problems rests on the commitment to a match between the explanation and the behavior (or behavioral component), that is, the commitment to preserve the explanation unit. The obligation to maintain the explanation/behavior match solves these three problems, because for every behavioral component, the researcher uses only one single explanation. (Note that commitment to the explanation unit is similar to commitment to a hypothesis: one adheres to it until it is disconfirmed.) As a result, the researcher cannot propose different explanations for the behavior under study, nor can she leap from explanation to explanation on a whim; she must use one explanation—the one determined as most suited to handling that kind of studied phenomena. This commitment prevents the possibility of proposing ad hoc explanations. The explanation unit is fixed in advance, as in the methodology applied in the natural sciences; there, for example, it is clear that the movement of bodies is treated by means of an explanation model of the kind proposed by Hempel (1965).

The commitment to the explanation/behavior match protects the theory from the problem of inconsistency. Since for every phenomenon, a matching explanation is determined in advance, no situation will arise where the researcher uses different explanations for the same behavior—that is, a situation of self-contradiction will not arise.

The explanation/behavior match also allows empirical and theoretical comparison of different theories. This comparison may be partial or entire—when the two specific theories based on the MEF use the same explanation procedures (schemes).

FINAL SECTION: OVERVIEW AND DISCUSSION

In this section I briefly review the present approach, compare the present approach with other relevant ones, and finally make a brief methodological remark.

The Present Approach

The major contribution of the present chapter is the proposal of a unique methodological approach, the MD, which leads to the development of the MEF. The latter helps a researcher to develop a specific psychological theory with an improved explanatory ability. The approach is not founded on the usual attempt to reduce mental states to neurophysiological states; it circumvents the ontological mind/body problem. The present approach is founded on a nontrivial integration of two kinds of explanation: mechanistic and mentalistic, which enhances the theory's ability to provide a better understanding of behavior than do the standard mechanistic theories developed by behaviorism and cognitive psychology. The present approach is established on the following major ideas:

a. Various specific purposive explanations are generated by the mentalistic scheme [Will/Belief]. Although [Will/Belief] cannot be reduced to a mechanistic explanation, it does fulfill the scientific methodological requirements for explanation.
b. Given (a), one may construct the MEF, which guides on how to develop a specific psychological theory based on two kinds of explanatory models: mechanistic and mentalistic (i.e., an MEF theory). The specific theory is

developed by making an appropriate match between behavior and the two kinds of explanation: the mechanistic or the mentalistic, called the explanation/behavior match.

c. The explanation/behavior match establishes the coherence of the specific MEF theory, and also offers solutions to certain problems raised by the proposed framework, which is founded on the multiplicity of explanation models.

Comparisons

Substance and property dualism: The present approach differs from substance dualism, which assumes that the mind is a different substance from the body, and also from property dualism, which assumes that the mind and the body are two different properties of the same substance. MD and MEF do not make assumptions of that kind, but bypass them and try to suggest a scientific methodology that unites mentalistic with mechanistic explanations.

Explanatory dualism: One may propose that the present approach is fairly close to explanatory dualism, which suggests that there is a fundamental difference between the explanation of nature and the explanation of action (e.g., Brook & Stainton, 2001; Maxwell, 2000; Sayre-McCord, 1989; see chapter 5). For example, Maxwell (2000) proposes that scientific explanation is not able to deal with conscious experience just as a blind man cannot grasp the essence of the experience of seeing the color red. To treat phenomena of this kind, use must be made of what he terms a "personalistic" explanation: a valid explanation, well-founded intellectually, and not given to reduction to a scientific explanation. Scientific and personalistic explanations are largely in conflict (when the former is understood, the latter becomes not understood, and vice versa), so the attempt to construct a super-theory uniting these two kinds is doomed to failure.

Although the present approach is close in spirit to explanatory dualism, it is distant from Maxwell's approach on several points. First, the mentalistic explanation applies to the individual's own behavior and others' behavior, including supreme animals' (e.g., they have conscious pain and fear and even suffer agony on separation). Secondly, arguments have been developed that show that mentalistic explanations are accepted by scientific methodology. Thirdly, the MEF was developed so that mentalistic and mechanistic explanations could be applied coherently to behavior.

Naturalism: The Naturalism approach suggests that mental phenomena, mentalistic behavior, should be investigated by a methodology that prevails in the natural sciences, explained by processes acceptable in the natural sciences, that are not semantic or intentional (e.g., Brook & Stainton, 2001; Polger,

2004; Von Eckardt, 1993). Even if one accepts that [Will/Belief] fulfills the demands of this methodology, the present approach can be viewed as naturalistic only in part since [Will/Belief] is a mentalistic explanation procedure.

Levels of explanation and multiple realization. Accordingly, behavior can be divided into three levels of description and explanation (see Block & Alston, 1984; Fodor, 1976, 1981; Pylyshyn, 1984; Von Eckardt, 1993):

1. the *mentalistic level*, which refers to private behaviors such as desire, belief, aims, intention, feelings, emotions, and consciousness;
2. the *information-processing level*, which refers to cognitive phenomena and cognitive concepts anchored to analogy with the computer, for example, information, information processing, storage, retrieval, coding, and computation;
3. the *neurophysiological level*, which refers to phenomena and concepts anchored to the domain of the natural sciences, such as physiology, neurology, and biochemistry.

For example, Von Eckardt (1993) writes: "The most important levels are the folk-psychology level, the information-processing level, and the neural level. It is also usually assumed that these levels are the products of (respectively) common sense, the non-neural cognitive sciences, and the neurosciences" (p. 318).

How are these three levels related? A common answer is that the connection is brought about by means of the realizability relationship: mentalistic processes are realized by processes of information processing; information processing is realized by neurophysiology. But in many cases, researchers speak of direct realization of mentalistic processes by neurophysiology.

On account of the multiple realization argument, it may be suggested that the functional explanation, the information-processing explanation for cognitive phenomena (perception, memory), is autonomous, and does not require a description that details how this explanation actually occurs physiologically (we may call it the "autonomous approach"). That is, explaining information processing is similar to explaining how the computer adds numbers by detailing the appropriate software without referring to the hardware processes (see Fodor, 1974; Putnam, 1967). For example, Von Eckardt (1993), who discusses the relation between the level of information processing and the neurophysiological level, supports the argument that she calls the "information processing ineliminability thesis" (pp. 330–339): it is not possible to eliminate the information-processing level because of the argument of multiple realizability, and because the neurophysiological level is not able to explain properly cognitive phenomena, which are well explained by the information-processing approach.

While these ideas are very important (and provocative) for the science of cognitive psychology, it should be emphasized here that by the present approach, the multiple realization argument for the autonomy of the information-processing level is not needed and can be circumvented. The main reason is that each of the above three levels (the mentalistic, the information-processing, and the neurophysiological) can be represented explanatorily by an appropriate and acceptable explanation model: the mental by a purposive account ([Will/Belief]), the information-processing by an appropriate mechanistic-computational account, and the neurophysiological by an appropriate mechanistic physiological account. The main problem of the present approach centers on the issue of making the proper match between explanation and behavior (and its components), namely, the explanation/behavior match.

There is an important difference between the autonomous approach and the present one. While the autonomous approach is satisfied with a computational explanation for a complex behavioral phenomenon, the MD and MEF approach evaluates this kind of account as insufficient. According to the latter, a better account should be proposed: an account based on both mechanistic and mentalistic explanations as integrated in an MEF theory.

Finally, the following remark should be made. An MEF theory is based on at least three kinds of language: mentalistic, information processing, and physiological. A similar phenomenon may be encountered in the sciences. For example, the explanation of how a flashlight works involves an account of the battery (a chemical mechanism) and the other parts (physical mechanisms). However, in this case, there are certain known transformations from language to language, which explain how the flashlight emits light. This is not the case with complex behavioral phenomena, since, as I repeat over and over again, a mind/body theory does not yet exist. Given this, one may ask: Can we develop a super-language that will express and unify the mentalistic and mechanistic languages? Well, I doubt it, but my official answer is that I don't know.

Methodological Remark

Although the [Will/Belief] explanation procedure meets the methodological requirements for giving a scientific explanation, the following methodological question may be raised: Does the [Will/Belief] procedure meet other methodological requirements for scientific research, such as the requirement that observation should be public? The immediate answer is that [Will/Belief] as an explanation procedure is not a kind of theoretical concept about which one may ask if it is publically observable or not. But this is not an exhaustive answer because the [Will/Belief] refers to the individual's inner world—will and belief, which are not subject to public observation. So the answer is more complex.

First, an argument may be raised that will and belief are not pure theoretical concepts and are not publically observable because only the person himself can observe his will and his belief as events that appear in his own consciousness. That is, apart from the individual himself, no other person is capable of observing the will and belief of that individual. Thus, these concepts are "concepts of restricted observation."

Secondly, as will and belief are concepts of restricted observation, it is possible to make a broad inductive generalization that every member of the human race experiences conscious events of the will and belief kind. Just as *X* can report a conscious observation of his will and of his belief, so can *Y* and *Z*, and just as the events will and belief appear in *X*, so do they appear in *Y* and *Z*. Furthermore, on the grounds of the behavior of supreme animals, and on the grounds of evolutionary theory, it can be assumed that such animals also have will and belief that direct their behavior.

Finally, on the basis of the foregoing, it can be proposed that the [Will/Belief] can be based on the will and belief concepts (as concepts of restricted observation), on the basis of which one may propose specific explanations for different behaviors that express one's will, belief, and behavior that realizes the individual's will.

Chapter 7

A Proposed General Explanation Procedure (GEP) for Understanding an Individual's Behavior

One of the most important goals of psychology as a scientific discipline is to provide understanding—understanding of behavior (see de Regt, Leonelli, & Eigner, 2009a; Nola & Sankey, 2007; see also chapter 1). To achieve this, a researcher uses many and varied research tools, for example, conducting observations, experimental research, statistical analysis, testing theories, giving explanations, and so on. This scientific activity is done in the framework of methodological rules (generally accepted by the scientific community) (e.g., Hanna, 1979; Kuhn, 1970).

Here I concentrate on procedures (schemes, models) for giving explanations that provide understanding by answering *why* and *how* questions (but see the discussion on whether understanding depends on explanation in Lipton, 2009, and chapter 1). I wish to offer a relatively new view of this subject in psychology.

Since its creation, psychology has undergone the most dramatic transformations: from research in consciousness to research in behavior, and to research in cognitive processes (e.g., Leahey, 2004; Rakover, 2012b). These changes were associated with changes in the procedures for giving explanations, for example, a change from explanations based on laws to those based on mechanisms (see Bechtel & Wright, 2009). In fact, psychologists have used a variety of approaches to provide explanations for behavior: some use internal mechanisms (mental, cognitive, neurophysiological); others infer (deductively, inductively) the studied behavior from a theory or a law, describe the behavior by a mathematical expression, or simulate it with an adequate computer program, and so on. This state of affairs arouses the following crucial questions: What is the most suitable explanatory procedure for individual behavior? Can one suggest a general explanation procedure that could be used to account for all kinds of individual behavior, or must one account for these behaviors by appeal to a multiplicity of explanation procedures?

The main object of the present chapter is to answer these questions. This will be carried out in the following way. The chapter will develop an explanation procedure (model) that will guide the researcher on how to suggest an account for an individual behavior. The procedure is tied to two anchors: (1) it is based on the actual ways psychologists propose explanations (in accordance with Bell, Raiffa, & Tversky, 1988, the present explanatory model was developed with an eye to reality for inspiration); and (2) the procedure is justified and established on sound and rational theoretical and philosophical arguments. (To the best of my knowledge, no such model that describes how psychologists suggest explanations for an individual behavior is to be found.)

I suggest here that in psychological research there is a general procedure for giving explanations for individual behavior, which I call the General Explanation Procedure (GEP). It is general since (1) it can handle different hypotheses, models, and theories (called "theoretical structures"), and (2) it can provide a theoretical framework for different models, schemas, procedures, and theories of explanations that have been discussed in the literature (see chapters 2 and 3). This procedure is characterized as not only logical (based on rules of deduction) but also methodological, as woven into the general scientific methodology for conducting research. That is, the GEP is viewed as an important tool among other research tools. Because the GEP is linked to the research tool of the theory-testing method common in psychology, I shall start with a very brief and schematic description of this method, and then go on to tackle the GEP.

Briefly, the theory-testing method recommends that the researcher take the following steps: (1) choose a theory (hypothesis, model, etc.) for theoretical and empirical reasons (detailed in the introduction to any research paper); (2) infer from this theory a prediction (description of the response of the participant in the research) given certain research conditions (stimuli); (3) run an experiment, a research project, meant to test the prediction empirically; and (4) check if the prediction matches the observation, the experiment's results, and draw one of two possible conclusions: when the prediction matches the observation, suggest that the given theory is confirmed; when the prediction does not match the observation, suggest that the theory is disconfirmed (see Haig, 2005; Rakover, 1990; Schweigert, 2006). This method is learned in introductory courses and is described in introductory books to psychology (e.g., Chow, 1987; Neale & Liebert, 1986; Rakover, 1990; Sternberg, 2009). Sternberg, for example, writes: "We usually start with a theory. . . . We seek to test a theory and thereby to see whether it has the power to predict certain aspects of the phenomena with which it deals. In other words, our thought process is, 'If our theory is correct, then whenever x occurs, outcome y should result'" (p. 13).

The theory-testing method is similar to the well-known Hypothetico-Deductive (H-D) method discussed in the philosophical literature (see Haig, 2005; Hempel, 1966; Nola & Sankey, 2007; Rakover, 1990; Sankey, 2008). For example, Hempel (1966) illustrates this method in his renowned book *Philosophy of Natural Science*. He describes the 1844–1848 research of Ignaz Semmelweis on childbed fever, which killed a large number of women giving birth in the Vienna General Hospital's first maternity division, but not those in its second. Semmelweis offers several hypotheses to explain this dreadful phenomenon. He tested them empirically by the H-D method, for which Hempel proposed this formulation (where *H* stands for hypothesis, *I* for prediction, and the horizontal line signifies a logical inference):

> "If H is true, then so is I.
> But (as the evidence shows) I is not true.
> --
> H is not true" (p. 7).

If the prediction does accord with the observation (the evidence), this does not mean that the hypothesis has obtained incontrovertible proof of its truth. It is usually said that the hypothesis is confirmed, or is supported by the observation.

Lipton (2005), who sharply criticized the H-D method, reanalyzed the childbed fever instance and argued that Semmelweis's research approach was more suited to Inference to the Best Explanation (IBE) than to the H-D method. (IBE is an inductive inference method where out of several hypotheses, the one that best explains the studied phenomenon is chosen. The name IBE was first proposed by Harman, 1965.) Although the H-D method has sustained much criticism highlighting its weaknesses, it is considered the most popular method in the sciences and in psychology (see Glymour, 1980; Haig, 2005).

From a review of the psychology literature and personal experience running experiments with animals and humans, I conclude that the H-D method is for psychologists also a basis for suggesting an explanation. This is not a straightforward conclusion since Salmon (1990) clarifies that the H-D method is not to be confused with Hempel's (1965) explanation model, the Deductive-Nomological (D-N) model. (On the D-N model, see chapter 2. To recapitulate, according to this model a law or a theory explains a phenomenon when the prediction, logically derived from the law or the theory and the experimental conditions, matches the description of the phenomenon under study.) Despite the similarity in the logical structure of these two research tools, their aims are entirely different. While the H-D method checks if the given theory is correct by means of prediction of the studied phenomenon—namely, the epistemological status of the theory is under investigation—the

D-N model offers an explanation for the phenomenon by means of a theory whose truth we assume—namely, the status of the theory is not under investigation. Rakover (1990), who made a similar distinction, suggested that the similar logical structure of the H-D method and the D-N model could serve two different purposes: to test for confirmation/disconfirmation of a theory, and to provide a basis for an explanation of the studied phenomenon by means of the tested theory.

In this chapter I expand the last suggestion as follows. Two results of applying the theory-testing method—(1) the prediction matches the observation; (2) the prediction does not match the observation—allow use of two research tools with two different functions of the studied theory:

- *The prediction matches the observation (the experiment's result)*: (a) by means of the theory-testing method it may be proposed that the tested theory has been confirmed, has obtained empirical support; (b) by means of the GEP it may be proposed that the tested theory supplies an explanation for the observation.
- *The prediction does not match the observation (the experiment's result)*: (a) by means of the theory-testing method it may be suggested that the tested theory is disconfirmed; (b) by means of the GEP it may be suggested that the tested theory does not supply an explanation for the observation.

In the case where the prediction matches the observation, psychologists turn to the GEP, and by its means suggest that the studied theory explains the observation under study (as well as other observations reviewed in the introduction to the research paper). In this case the tested theory answers *why* and *how* questions concerning the experiment's findings. Since the answers to these two questions (why and how) are conceived by many (e.g., Lambert & Brittan, 1992) as important components in the concept of scientific explanation, it can be suggested that GEP is a procedure for giving explanations.

But if the prediction does not match the observation, psychologists tend to say that the theory has been disconfirmed. In this case one does not use the refuted theory for explanation, simply because it is incorrect. Given this, the following should be noted: it is easy to conceive that an incorrect theory may produce a suitable prediction (matching the observation), but it is hard to conceive of a correct theory producing a poor prediction (and see below on the relation of explanation to prediction).

It is very important to stress that a theory in itself cannot give an explanation for the studied phenomenon because in essence it is an idealized description of certain regularities, rational relations among certain variables (and here I avoid elaborating the concept of scientific theory; see Suppe, 1977b). For a theory to be able to function explanatorily, it must be part of a procedure designed to give explanations, such as the GEP. In other words, the capacity of the theory to offer explanations is realized by the appropriate procedure, a procedure for scientific explanation. Similarly, the capacity of the theory to pass an empirical test is realized by the proper procedure, namely, the theory-testing method. These two procedures thus realize two important theory dispositions: to propose explanations and to pass an empirical test.

The GEP is based on three main components. Two of them belong to the theory-testing method: the tested theory and the relation obtained between the prediction and the observation. The third component, which turns it into an explanatory research tool, is the assumption about the way the tested theory provides an explanation, how the explanation of the studied phenomenon is provided by the theory under study (see below).

The chapter is organized in the following way. The first section discusses the question, "What is the General Explanation Procedure?" The next concentrates on five theoretical and philosophical problems underlying the GEP. The third deals with the generality properties of the GEP. And finally, the discussion section treats several broad philosophical ideas important for the GEP.

WHAT IS THE GENERAL EXPLANATION PROCEDURE?

The GEP is based on certain rational processes, which occur on the following three conceptual levels (see Figure 7.1).

Observation level (II) deals with the observations made on the individual who is in a stimulus state and responds with a certain response. For example, in a face recognition experiment the participant decides whether the presented face is familiar or not.

Observation level (I) *assumes* that there is in the individual some Unknown Real-Process (URP) that is responsible for giving his/her responses; namely, it is *assumed* that *Response* = *f*(*Stimulus*, *Unknown Real-Process*). (I use the word *responsible* to roughly cover all kinds of explanatory relations, such as causality, statistical, mathematical, and logical.) Despite this basic assumption's being lodged in the absence of knowledge (ignorance of the nature and

Explanation Level: *{Stimuli, Theoretical-Structure}* ⟶ *Response*

Observation Level (I): (Stimuli, Unknown Real-Process) ----➤ Response

Observation Level (II): (Stimuli, Individual) -·-·-·-·-·-·-➤ Response

Figure 7.1 The Three Conceptual Levels of the GEP.

structure of this process, how it operates, etc.), this is an extremely important assumption because it is what makes the GEP a procedure for giving explanations.

Explanation level aims at the assumption of the URP. It attempts to overcome ignorance regarding the URP by presenting relevant "theoretical structures" (hypotheses, models, mechanism, theories, etc.) that describe as accurately as possible the nature and the operation of the URP in the function *Response* = *f*(*Stimulus, Unknown Real-Process*). The explanation level, then, presents theoretical structures able to answer *why* and *how* questions connected to the behavioral phenomenon under study. It should be emphasized here that except for the essential assumption that there is a URP, all its possible attributes are proposed by the theoretical structures. It is the latter concept that suggests whether the URP is a certain causal factor (necessary or sufficient) or a causal process; whether it can be described by statistical processes, evolutionary processes, or cognitive, mental, or neurophysiological processes; whether the URP's activity can be captured by logical processes (deductive, inductive) or mathematical or practical considerations; and whether the URP can be approached by a functional relation among stimulus–response–consequence without any reference to mental processes.

Many philosophers have assumed that there is something (unknown, of course)—which is referred to as a reason, argument, ground, function, cause, mechanism, process, and so on—that carries the account for the phenomenon under study. It seems that underlying numerous explanatory models, there is a kind of Unknown Real-Process on which explanation is based (see e.g., Salmon, 1990; Woodward, 2011; chapters 2 and 3). For example, Salmon (1984) proposed that the world rests on a causal structure, so he suggests that to give scientific explanation is to show how the studied phenomenon fits into the causal structure of the world. I will not confine myself to the approach that explanation is founded only on causality (since there are other forms of explanation), but I will base my proposal for an explanation procedure on the assumption that there is a very broad and general URP.

Given the above, explanation can be delineated in the following way: explanation depends on a theory, which has passed the empirical tests, and is assumed to approximate the URP that is responsible for the occurrence of the studied behavioral phenomenon. In other words, the studied phenomenon is explained by the theory, because the theory approximates the URP, which in reality is responsible for the occurrence of that phenomenon.

To illustrate the great importance of the assumption of a URP, let us look at the following fictitious case. Assume that an advanced electronics lab has succeeded in building an intelligent robot called Max. Now also assume that after a great length of time (say about 2,500 years) researchers find Max in an abandoned junkyard and begin to study this strange creature. They show it a series of stimuli and obtain a series of responses accordingly. Now these researchers try to construct a theory whereby they will be able to offer an explanation for Max's behavior. The researchers assume quite naturally that there exists in Max some sort of URP responsible for its responses in certain stimulus states, and that the theory they are developing is an approximation to this Real-Process. Why would these researchers assume (consciously or not) the existence of a URP? For several reasons. First, Max's responses were not fortuitous but evinced order (albeit not a simple order, because Max, as noted, is an intelligent robot), and order in behavior naturally leads to an assumption about some URP responsible for the behavior. Secondly, the researchers got the impression that this real-process is present in Max and not outside it, because Max did not respond in any of the cases where its sensors did not receive the stimuli presented to it by the researchers. Thirdly, several researchers (historians of science) discovered that even in the remote past, ancient science had made considerable achievements, so to understand Max the mechanical-electronic mechanism (which had been developed at that time) responsible for its behavior had to be discovered. Fourthly, some of the researchers focused on the analogy between the assumption about a URP and a physician's assumption (then and now) that if the patient's behavior deviates from the norm, then apparently something in that patient's system has gone wrong, and the physician's task is to uncover the URP responsible for the illness, and to prescribe the proper treatment.

Although the assumption about the URP seems perfectly natural, we still must ask why this assumption is vital. Why is it impossible to use the theory-testing method itself, and state that in the case of successful prediction, the studied theory explains the result of the experiment, the observation? The answer, quite simply, is that the D-H method lacks the methodological component for giving an explanation, and that this component finds expression in the GEP—the assumption of a URP. Any procedure, scheme, or model of explanation contains a component characteristic of an explanation, a component that makes the procedure a supplier of an explanation. For example, according to the D-N model, the explanation is conceived as a logical argument: the explanation of the studied phenomenon is given when from a law and certain experimental conditions one deduces a conclusion that matches the description of the phenomenon. According to Salmon's (1984, 1990) Causal-Mechanical model, the explanation of the studied phenomenon is given through a description of the causal process propagating in space and time, and which leads to the given phenomenon. According to the information-processing model analogous to a computer, the explanation for behavior is given by a description of the activity of cognitive components operating on units of information, which produces the studied behavior. In the present case of the GEP, the explanation for the given behavior is made possible through the URP assumption, which directs the activity of the explanation level to propose a theoretical approximation to this process, which answers *why* and *how* questions.

So the explanation level attempts to simulate what happens in the individual in a stimulus state. Given the assumption of a URP, the explanation level suggests a possible decoding of this process: a theoretical structure, which is based on rational processes so as to produce a description of the response. In Figure 7.1, the broken dotted arrow on observation level (II) describes the response as dependent on the stimulus state and on the individual. The broken arrow on observation level (I) describes the response as depending on the stimulus state and on a URP (which is part of the intricate system that composes the individual). The solid arrow on the explanation level describes the *response* as dependent on the *stimulus state* and on the *theoretical structure* proposed as mirroring the URP.

In light of the foregoing, it may be suggested that the GEP is based on the following *Explanatory Characterization*: if it is possible to rationally produce from the theoretical structure a successful prediction, we may then say that the theoretical structure offers an explanation for the phenomenon under study. This is justified by the following consideration: (a) It may be suggested that a real-explanation is given to the studied behavior by the URP, since the real-process is responsible for the individual's behavior in a certain stimulus state—that is, the real-explanation is anchored to an assumption that *Response* = *f*(*Stimulus*, *Unknown Real-Process*). (b) On the assumption that the theoretical structure is an approximation to the URP, one may suggest the following analogy: just as the URP suggests a real-explanation for the phenomenon under study by virtue of its being responsible for the response, so the theoretical structure offers an explanation for this response by virtue of its being an approximation of the *Unknown Real-Process*.

If so, in this case, where the predicted *response* matches the individual's response (see Figure 7.1), it may be suggested that the occurrence of the individual's response becomes understood by means of the theoretical structure that provides it with an explanation. This understanding, however, is not absolute but provisional: first because other theoretical structures may predict the same response given the same stimulus state, and secondly because further research may discover that under new conditions the given theoretical structure does not succeed in generating good predictions. In other words, the explanatory capacity of the given theoretical structure depends on the development of research, so one may suggest that this capacity is not fixed; every explanation is provisional, and in truth scientific research never ends. In these respects, I wholly agree with Popper (1995), who proposed the following methodological rule: "The game of science is, in principle, without end. He, who decides one day that scientific statements do not call for further test and that they can be regarded as finally verified, retires from the game" (p. 53). (See also Einstein & Infeld, 1938/1950.)

The present procedure deals with two kinds of explanation: "explanation" offered by the theoretical structure, and "Real-Explanation" offered by the URP. The difference between explanation and Real-Explanation is somewhat similar to that between actual explanation and potential explanation (see Hempel, 1965; Lipton, 2005). Actual explanation is the correct explanation for the given phenomenon, since it meets the requirement that its *explanans* must be true. For example, in an actual D-N model the explainers, the law and the experimental conditions must be true. The potential explanation does not set this requirement. Potential explanations can be alternative explanations to the studied phenomenon. In this respect the explanation proposed by the theoretical structure may be seen as providing a potential explanation. We do not know if the explanation is identical to a real-explanation proposed by the Unknown Real-Process; all we know is that the theoretical structure suggested a successful prediction, so its explanation is closer to a real-explanation than are failed explanations suggested by other theoretical structures.

In this light, an important difference may be noted between the present approach, which posits an explanation/real-explanation distinction, and the approach that posits a distinction of a potential/actual explanation. While the former assumes the existence of a real-process in the individual that may be approximated by explanatory suggestions on the explanation level, the latter distinguishes the two kinds of explanation, potential/actual, theoretically—a distinction that rests on fulfilling the requirement of truthfulness of the *explanans*. This requirement gave rise to an insoluble question in the philosophical literature: What distinguishes a law from an accidental generalization? Apart from simple cases where it is obvious that no universal law is involved, but an accidental generalization (e.g., the "law" holding that all the guests at Ruth's party last Friday evening had high intelligence, IQ above 150), it is hard to suggest any criterion in this matter. In special sciences such as psychology, this question is particularly pronounced: Do we have before us a law, an accidental generalization, or perhaps a local phenomenon that requires an explanation? This problem is bypassed by the GEP because the explanation on the explanation level is conceived as a provisional approximation to a real-explanation. This approximation, as stated, is based on the research objective to discover the nature of the URP, a goal that according to Popper will never be reached. In other words, the explanation suggested by the GEP is not dependent on the truthfulness of the *explanans* (especially the law), since the explanation according to the GEP is founded on approaching the URP.

FIVE PROBLEMS UNDERLYING GEP

I discuss five questions that form the theoretical underpinning of the GEP: (1) What are the relations between different explanation procedures and different rational processes? (2) Does the division between the explanation level and the observation level hold? (3) Is GEP in its essence a realistic procedure? (4) What is the relation between explanation and prediction? (5) How can the assumption that the theoretical structure approximates an Unknown Real-Process be justified and evaluated? These five questions touch on a wide and thorny array of topics in philosophy, which have been constantly debated over a long period. Therefore, I shall consider them briefly, and propose answers that are only relevant to GEP.

Rationality

It is accepted that the rules of logic, probability, statistics, and the theory of decision-making are normative rational systems. Yet it turns out that often people do not act according to these rational norms. The explanation for deviations from the norm sparked debate: on the one hand, the heuristics approach suggests that people act according to automatic heuristic systems that for certain problems create biases in judgment, decisions, and logical considerations; on the other hand, these deviations can be explained by practical limitations of the individual, environmental constraints, difficulties in applying rational normative rules to the studied problems, and also by some other rational rules arising from the individual's background and the situation to which he is subjected (see Gigerenzer & Goldstein, 1996; Kahneman, 2011; Shafir & LeBoeuf, 2002; Simon, 1991; Sturm, 2012; Tversky & Kahneman, 1974).

Although in a given research field, and certainly in different research fields, rational behavior is often characterized in different ways, it may be suggested that in general, *rationality* refers to the goals and to actions most suited to realizing these goals, while the constraints of reality are taken into consideration (e.g., Ellis, 2008; Hempel, 2001; Newell, 1982; Popper, 1994). For example, Popper (1994) describes the "rationality principle" as the way in which one acts according to the situation, including knowledge and means of realizing one's goals. Newell (1982) suggests the principle of rationality: "If an agent has knowledge that one of its actions will lead to one of its goals, then the agent will select that action" (p. 8). Ellis (2008) posits the basic idea of instrumental rationality thus: "Instrumental rationality picks out actions that best satisfy each (set of) end(s) in each situation (at least as it is perceived). In doing so, it defines a correspondence between ends and situations on the one hand and actions on the other" (p. 202). Hempel (2001), dealing with the rationality of sciences, suggests this:

> To put the point somewhat loosely: a mode of procedure is rational, relative to a certain goal and a given of means-ends information, if, judged by that information, the procedure offers an optimal chance of attaining the goal. . . . In regard to the procedural rules laid down in laboratory manuals for the pursuit of certain limited and highly specific objectives, such as the measurement of particular quantities or the experimental testing of particular kinds of hypotheses, it can quite plausibly be argued that—given the current scientific knowledge in the field—they qualify as optimal and thus as rational in the sense indicated. (p. 362)

In light of the foregoing, I shall attempt to show that different rational rules form the basis for different procedures, schemes, and models of explanation. While the purpose of the scientific explanation—to offer understanding by answering *why* and *how* questions—is agreed by most researchers, the rational ways of realizing this purpose are many and varied. First I shall enumerate several important schemes and models of explanation in the sciences discussed in the professional philosophical literature that are based on different rational rules (see chapters 2 and 3); then I shall describe the application of some of them in psychological research.

In addition to the Deductive-Nomological (D-N) model, Hempel (1965) offered two more models to explain probability phenomena: the Deductive-Statistical (D-S) model, which is very similar to the D-N model but includes in its premises a statistical law and predicts deductively the probability of the occurrence of the given phenomenon; and the Inductive-Statistical (I-S) model, which is like the D-S model but includes in its premises a statistical generalization and predicts inductively the probability of the occurrence of the phenomenon. Following widespread criticism of Hempel's approach and an extensive discussion of the explanation concept, several alternative explanation models were proposed for various research fields, and these too encountered much criticism (e.g., Gijsbers, 2007). These alternatives include the Statistical-Relevance (S-R) model (Salmon, 1971, 1990), the Causal-Mechanical model (Salmon, 1984, 1990), the Unificationist model (Kitcher, 1989), the Manipulationist-Interventionist model (Woodward, 2003), the Pragmatic model (van Fraassen, 1980), and explanation by mechanism (Bechtel, 2008a, 2008c; Craver, 2001). These explanatory models are based on various rational rules; for example, logic rules form the basis for the D-N model, inductive rules for the I-S model, statistical rules for the S-R model, causal processes for the Causal-Mechanical model, and different kinds of mechanisms (including information processing) for explanation by mechanism.

Some of these explanatory models have been applied in psychology (see Thagard, 2012). The application encounters a serious problem: laws or top theories have not yet been discovered in psychology, as they have in the sciences (especially physics) (e.g., Rakover 2012b). So it is hard to transfer to psychology the explanatory schemas based on such laws or top theories. Cummins (2000) suggests that it is impossible to apply the Hempel approach to psychology because the laws discovered in psychology are no more than empirical phenomena that themselves require explanation; therefore, they cannot function as explanatory components in the D-N model. He maintains that in psychology, functional analysis should be applied, which supplies explanation by means of a mechanism. Likewise, Bechtel and Wright (2009) state that explanations in psychology are offered not by use of laws but by mechanistic explanation, namely, explanation based on description of a mechanism (see chapter 3).

Thagard (2012) proposes that in psychology four approaches are taken to build explanatory models: deductive models (e.g., the D-N model) or systems based on "if . . . then" rules, which are based on rules of deduction; analogous schemas based on a match of the *explanans* and the *explanandum*; probabilistic models (e.g., the Bayesian models); and neural network models based on activation and connectivity processes. Simon (1992) suggests that in psychology a new way has arisen for giving explanations for complex behaviors: explanations based on a number of levels, symbolic and physiological, that are complementary and not competitive. Simon emphasizes the great importance of explanation by means of theories that are computer programs (simulations) whereby complex behaviors become explicable, when the symbolic level (the information-processing program) is essentially independent of the neurophysiological level. Here it is worth noting that several researchers propose three levels of description and explanation (see discussion in Rakover, 2007): (1) the *mental level*, for explaining behavior through mental concepts such as will and belief; (2) the *information-processing level* for explaining behavior through cognitive concepts and processes in analogy to the computer; and (3) the *neurophysiological level*, which explains behavior through neurophysiological processes in the brain. Based on the well-known argument of multiple realizations, one may refer to each of these levels separately and suggest a description and explanation of phenomena on the particular level without reference to the other levels.

Theoretical and Observational Concepts

The classic approach to a theory structure, based on the positivistic approach to science, posited, among other things, a distinction between an observational concept and a theoretical (hypothetical, unobservable) concept, in order to base scientific knowledge (e.g., laws, theories, models) on a solid observational and objective infrastructure. However, this distinction was subjected to severe criticism, which led, on the one hand, to different developments in the structure of a scientific theory, and, on the other hand, to an ongoing debate between those who rejected and those who accepted this distinction (see Bogen, 2013; Clark & Pavio, 1989; Lambert & Brittan, 1992; Rakover, 1990). One of the arguments against this distinction held that observational concepts are not pure because they are influenced by various theoretical approaches—that is, observational concepts are theory laden. For example, Hanson (1958) posed this interesting question: Did Kepler and Tycho Brahe see sunset in the same way? Hanson replied in the negative. Because Kepler believed in the heliocentric approach, he saw a horizon moving against a fixed sun; by contrast, Tycho, who believed in the geocentric approach, saw the sun moving against the background of a static horizon. Furthermore, it arose that even almost self-evident observational concepts, such as distance, were laden with theoretical meaning, as the theory of measurement shows (see Michell, 1999). Even so simple a concept as the percentage of correct responses is not free of the effect of background and theoretical considerations, as it becomes clear from Signal Detection Theory (STD) (see MacMillan & Creelman, 2005).

My position on this matter is that despite the important criticism of this distinction, it has not yet received its coup de grâce: it is of great importance theoretically and practically, and psychologists use it still today. For example, it is hard to find a psychologist who confuses concept group (I)—reaction time, pressing on a pedal, eye movement, heartbeat, and breathing—with concept group (II)—ego, instinct, visual scheme, consciousness, perception, and long-term storage—and will suggest that group (I) contains theoretical concepts and group (II) observational concepts; or say that there is no difference between the two groups. Clark and Pavio (1989), who conducted empirical studies on this issue (grading seventy-two psychological concepts on different measures such as observation and abstract), would probably concur with this statement. In their paper, they summarize the theoretical–empirical discussion thus:

> The observational-theoretical distinction is supported by empirical tests of its validity. . . . Moreover, rational arguments against the distinction are inconclusive, and other phenomena of science suggest that the distinction is generally valid. . . . Scientists do and ought to maintain distinct attitudes toward observational and theoretical terms when thinking or communicating scientific ideas. (p. 510)

Here it is worth adding two important comments with significant implications for the triple division I suggested above, namely, explanation level, observation level (I), and observation level (II). The first comment is that even if we accept that observational concepts are laden with theoretical meaning, this does not constitute a stumbling block to scientific development. The theoretical meaning of a tested theory is different from the theoretical meaning imparted to the observational index against which that theory is tested empirically. For example, in most cases SDT does not relate to the studied theory, and the use of indexes such as the parametric d' or the non-parametric A', which according to SDT are a correction and improvement on the %C (percent correct) index, does not affect the tested theory itself. It is a fact that many and varied hypotheses and theories in psychology (attention, perception, memory, etc.) have been tested empirically by means of these improved observational indexes. Similarly, one may propose that all the sophisticated instruments (e.g., Polygraph, EEG, fMRI) used by psychologists in their experiments are anchored to different theories. This fact does not impede the use of these instruments to obtain a special kind of an individual's responses, namely, empirical indexes, by means of which it is possible empirically to test many and varied theories in diverse experimental conditions.

The second comment is that the very division into three levels—one theoretical and two observational—is theoretical in essence. This division proposes that the explanation level is clearly based on theoretical concepts, but the two observation levels also are linked to theoretical conceptualization, albeit of different degree: observation level (I) is more heavily laden with theoretical conceptualization than observation level (II) because of the assumption about a URP. So intuitively, these three levels might be perceived as three points on an ordinal scale of theoretical/observational. The question is, How shall we justify this intuition? To answer, I suggest the following considerations.

First, can the assumption of a URP in itself be used to suggest a satisfactory explanation for the individual's behavior? My answer is negative. That assumption is located on a level between the explanation level and observation level (II), because it is an extremely general assumption: there is something in the individual—a URP—that is responsible for his/her behavior. As such, it cannot provide a specific prediction, simply because we do not know what the nature of a URP is. To suggest specific predictions, we must turn to the explanation level, which will provide a theoretical structure that will offer such relevant predictions regarding the behavior under study.

Second, should not the assumption about a URP be included on the explanation level? Since this assumption is theoretical, it may be suggested that its place is indeed on that level. Again, my answer is negative. The main reason for this is that the goal of the explanation level is to decode the URP by a proposal for a theoretical structure. As a goal to which the action of the explanation level is directed, the URP resides outside the explanation level. Similarly, the finish line of a sprinter lies one hundred meters distant from him.

It is thus possible to suggest that, on the one hand, observation level (I) is a level that touches on observation level (II) (because the URP is an assumption about something that exists in the individual), and on the other hand, it is a level that touches on the explanation level (because the assumption about the real-process being responsible for behavior is theoretical).

Realism

The distinction between realistic theories (interpreted as proposing proximities to what takes place in the world, which is conceived as existing independently of the individual's mind, including others' feelings and thoughts) and instrumental theories (interpreted as only theoretical and computational means to predict empirical phenomena) is linked to the foregoing distinction between theoretical and observational concepts. The realistic approach suggests that theoretical concepts represent a description approximate to what takes place in the world, while the instrumentalist approach suggests that theoretical concepts are nothing more than means of calculating a good prediction of the observations (see Chakravartty, 2013; Papineau, 1999; Nola & Sankey, 2007; Rakover, 1990). In these respects, it may be suggested that the GEP is connected to the realistic approach. The theoretical structures on the explanation level are not concepts bearing an instrumental function alone (efficient concepts for generating predictions) but are structures that attempt to reflect what takes place in the individual. This reflection, as stated, forms the basis for justifying the statement that the GEP is a procedure for giving explanations approximating the real-explanations, which are provided by the URP. (I shall resume the matter of realism in the "Discussion" section because of its great importance for the GEP.)

A large part of cognitive psychology may be interpreted as based on the realistic approach, since the explanation of the individual's behavior is based on the analogy to a computer: the cognitive processes are analogous to the software, and the neurophysiology of the brain is analogous to the hardware. By this analogy, an individual's cognitive behavior is explained by rules that act on physical symbols representing the environment, namely, through information processing. This is not only a realistic description of the cognitive system, but also, as several researchers posit, a calculating process that the mind puts into practice (e.g., Thagard, 2008). Here are several examples (from visual memory and face perception) that support this interpretation. Sperling's (1960) famous experiment was an ingenious variation on earlier experiments, which found that the capacity of the attention span was limited to about four items of the information presented to the participants (e.g., a $3 \times 4 = 12$ matrix of letters and numbers). But Sperling found that the whole information was stored in the visual memory for a short time (less than a second). Rakover and Cahlon (1989, 1999, 2008) developed the Catch model as a system operating in the individual in order to explain how a face may be reconstructed from a witness's memory; Rakover and Cahlon (2013) developed the Face-Checking model as a system acting in the individual in order to explain the perception of scrambled faces; and Rakover (2013) developed the Face-Scheme Incompatibility (FSI) model as a system at work in the individual in order to explain the Face-Inversion Effect (FIE). In many more cases, researchers who believe that face perception is special and innate have supported their theories by the neurophysiological discovery that there exists in the brain a special area that handles face perception—the facial area, which is activated by facial stimuli (e.g., Kanwisher, McDermott, & Chun, 1997; McKone & Robbins, 2011).

Explanation and Prediction

As evident from the discussion so far, the terms of explanation and prediction are fundamental to the GEP. What are their methodological relations? On the one hand, a successful prediction (a description that accords with the individual's behavior) may support the supposition that the theoretical structure provides an explanation for this behavior by virtue of its being a reflection of the URP responsible for the individual's behavior. On the other hand, this does not mean that a successful prediction is itself decisive proof that the theoretical structure is indeed a mirror image of what takes place in the individual. To substantiate this difference between explanation and prediction, we may take another look at the fictitious example of Max the robot.

Assume that the electronics lab that built Max was able to program it in such a way that it responded to type-*S* stimuli with type-*R* response, so that $R = f(S)$ showed an inverted *U* function—that is, to weak or strong *S* Max responded with weak *R*, while to medium *S* Max responded with strong *R* (see Figure 7.2, function A). Now, if one asked why Max responded to strong *S* with weak *R*, one would get the following real-explanation: Max is so constructed that the relation of *S* to *R* upholds the inverted *U* function, so Max responds to strong *S* with weak *R*. (The answer to the question why Max was programmed that way is an entirely different story.) So much for the real description of Max the robot.

Now let us further suppose that after some considerable time (as noted, 2,500 years) two scientists investigated Max separately. Scientist 1, who managed to activate Max, showed it several groups of type-*S* stimuli, and accordingly obtained several type-*R* responses. He submitted the data to a powerful statistical program and found that theory B (function B in Figure 7.2) furnished a good explanation for Max's behavior. Scientist 2 was able to present Max with only two groups of type-*S* stimuli, and found type-*R*

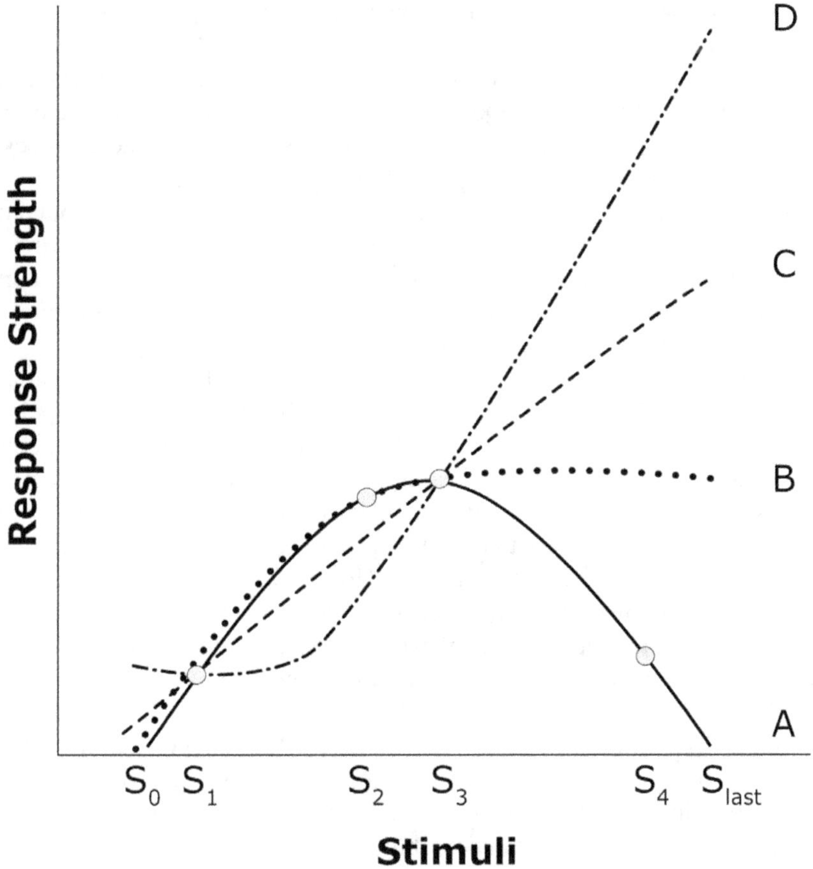

Figure 7.2 Four Theories that Explain Marx's Response Strength as a Function of Stimuli.

responses. Then Max went irreparably wrong. This scientist too submitted her data to a statistical program, and found that theory C (function C) furnished a fitting explanation for Max's responses. Thus it emerges that two observations may be predicted—a(S_1R_1) and b(S_3R_3)—by means of both theory B and theory C. (The second scientist also thought about theory D, but for simplicity's sake she preferred theory C.) So here the question is, Can we take it that both these successful predictions indicate that these two theories reflect real processes relevant to Max's behavior? Obviously we cannot (because Max was built on the basis of the inverted *U* theory, function A in Figure 7.2). But because neither scientist knows anything about function A, the interesting question is: What does scientist 2 think? The answer is that she will assume on the basis of the two good predictions that the linear theory C reflects relevant processes in Max, and that this theory offers a good explanation for both Max's responses in the particular stimulus states. However, her assumption and her explanation are only provisional. Why? Because if scientist 2 reads scientist 1's research report she will understand at once that the linear theory has failed, while scientist 1's theory, B, seems like a more successful one—that is, the predictions support the supposition that this theory suggests a better approximation to the real-process and to the real-explanation. (Clearly, scientist 1's theory is not correct either. But it is worth noting that with knowledge of the true theory of the inverted *U* function, for a certain observation range—S_0 to S_3 [see Figure 7.2]—scientist 1's theory may well offer a nice approximation to the real-process.)

In the philosophy of science, the connection between explanation and prediction has undergone interesting changes. First, Hempel and Oppenheim (1948) proposed the D-N model and asserted that explanation and prediction stem from the same logical structure (the difference between them is conceived as dealing with the occurrence of events in time: explanation is intended for events that happened in the past and for prediction of events that will happen in the future). However, this approach, called the symmetry thesis, was subjected to such severe criticism that the deductive model was abandoned; the term *explanation* assumed a central place at the expense of *prediction*, and the literature was flooded with alternative models of explanation (see Bechtel, 2008c; Douglas, 2009; Rakover, 1990; Papineau, 1999; Salmon, 1990; Woodward, 2011; chapters 2 and 3). The following instances show that prediction is not always explanation: a storm may be predicted from a fall in the barometer, yet the explanation for the storm does not lie in the barometer (but in atmospheric pressure); the height of a volcano may be predicted from knowledge of the length of its shadow and the angle of the sun's rays, but the explanation for the volcano's height does not lie in the shadow's length and the angle of the sun's rays (but in internal volcanic processes inside Earth); although Newtonian theory was able to propose very successful predictions about heavenly and earthly events that were ascribed to gravitational force, Newton could not explain this force (see Hon, 2001).

The following example shows that explanation does not suggest a prediction in every case. Darwin's theory, natural selection, offers a good explanation for the appearance of different species, but it is hard to predict evolutionary development on the basis of this theory (e.g., Douglas, 2009; Naylor & Handford, 1985; Scriven, 1959). Naylor and Handford (1985) write that one critique of the Darwinian theory (a critique that these authors do not agree with) is this: "The theory of natural selection lacks the predictive aspect needed to test it by the Popperian ideal of falsification, and is therefore not truly scientific" (p. 478).

Naylor and Handford's (1985) article, as well as Douglas's (2009) publication, reject the criticism that Darwinian theory is unable to make predictions that are empirically testable. Naylor and Handford discuss a considerable number of instances that support this rejection, for example, the discovery of mammal fossils in Antarctica; changes of color in a moth population due to industrial dirt that made the moth easy prey; the development of resistance to DDT in insects; and construction of various models to deal with diverse evolutionary developments. So it seems that the present approach to the relation of prediction to explanation was not undermined by the example of evolution, and accords with that of Douglas (2009): "In other words, the scientific nature of an explanation is to be found in its ability to generate testable predictions" (p. 445). "It is the ability of an explanation to generate new predictions, which then serve as a check on the explanation that improves the accuracy of our scientific explanations" (p. 454). Finally, as additional thought-provoking support for the immense importance of prediction in the development of scientific theory, consider what the Nobel laureate for physics, Feynman (1985), writes for readers of his book *QED: The Strange Theory of Light and Matter*:

> It is not a question of whether a theory is philosophically delightful, or easy to understand, or perfectly reasonable from the point of view of common sense. The theory of quantum electrodynamics describes Nature as absurd from the point of view of common sense. And it agrees fully with experiment. So I hope you can accept Nature as she is—absurd. (p. 10)

Approximation to the Unknown Real-Process

The question of approximation of the theoretical structure to the Unknown Real-Process may be conceived in the framework of the broader question of the degree of proximity of scientific theory to the truth (to the theory or law, which describes true occurrences in the world). According to Popper (1963), because every scientific theory will in the end be found incorrect, the question arises as to what the degree of proximity of an incorrect theory to the truth is. This question is known in the professional literature as the problem of "truthlikeness" or "verisimilitude." The history of science is replete with cases where an incorrect theory is replaced by another theory, which eventually also proves incorrect. This replacement of theories may well be a reason for pessimism since even the latest theories are liable to be found wrong, or for cautious optimism, which sees this replacement as taking us ever closer to the truth (see Thagard, 2007). Personally, I tend to see the replacement of theories as a process that in the long term brings about scientific progress, like improvement of behavior by learning from mistakes.

Let us look at a simple case as an example of approximation to the truth (following Oddie, 2007): assume that theory A posits that the solar system has five planets, while theory B posits fifty planets. Both theories are incorrect, but intuitively it seems that theory A is closer to the truth (eight plants) than theory B. To solve the problem of proximity to the truth, Popper put forward an idea that I shall describe briefly: he proposed seeing a theory as composed of two kinds of information—true and false—and determined that theory A is closer to the truth than theory B when the true information of theory A is greater than that of theory B, and when the false information of theory A is less than that of theory B. This notion of grading (potentially or actually) incorrect theories by degree of proximity to the truth has been subjected to fierce criticism, resulting in alternative ideas, which themselves have given rise to sharp critique and persistent debate, which I cannot deal with here (see Nola & Sankey, 2007; Oddie, 2007; Psillos, 1999; Rakover, 1990). The question that arises in this connection is: What is the yardstick for deciding if theoretical structure A is closer to a URP than theoretical structure B?

The solution I propose is linked to the basic idea of the GEP: we do not know whether the theoretical structure is identical to the URP, but we posit that theoretical structure A, which offers a successful prediction, is closer to the URP than theoretical structure B, which offers a failed prediction. In other words, if the theoretical structure approximates the Unknown Real-Process, then the behavior under study can be given an explanation by this theoretical structure. On this basis I propose a "Proximity index," which grades different theoretical structures according to proximity to a URP. To substantiate this criterion, we shall yet again take a look at Max and its responses, as these are depicted in Figure 7.2.

Recall that in Figure 7.2, four different theories appear, expressed for simplicity as four different functions describing the relation of strength of response to the stimuli, where the observation range is from S_0 to S_{last}. Theory A, namely the real-process by which Max was built, predicts perfectly the strength of the response as a function of the stimuli in the S_0–S_{last} range. Theory B predicts Max's responses well across the S_0–S_3 range. Theory C and theory D predict well Max's responses to the two stimuli S_1 and S_3. As may be seen from Figure 7.2, the three theories B, C, and D are incorrect. Nevertheless, it is intuitively clear that theory B is more proximate to a real-process (to theory A, by which Max was built) than is theory C, which is closer to a real-process than is theory D. On the grounds of this example, I propose the Proximity index, as based on the following rules:

1. *The theories concerned with the studied behavior must allow predictions in the relevant observation range.* As may be seen from Figure 7.2, the observation range is S_0 to S_{last}. The four theories allow predictions for each

of the stimuli within these limits. If a theory is concerned with only part of the observation range, or if it also concerns irrelevant observations, the extent of its possible approximation to a real-process is reduced from the start, because it will not be able to explain all the relevant phenomena or deal with phenomena that are not of the kind under consideration.

2. *The Proximity index stands in inverse relation to the error in prediction: the lower the index score, the greater the proximity to the Unknown Real-Process.* A possible calculation for the Proximity index is made by summing the squares of errors over the relevant stimulus states $\sum(P - O)^2$, where P is the prediction of the response and O is the observation of the individual's response. (This calculation accounts for the possibility that P is greater or smaller than O.) As Figure 7.2 shows, over the four stimulus states—S_1, S_2, S_3, S_4—the sum of errors^2 of theory A is zero, while the sum of errors^2 of the three incorrect theories grades them according to their degree of proximity to the real-process thus: theory B, theory C, and the farthest from the real-process, theory D. (Above a certain $P - O$ size, it may be said that the given theory is disconfirmed.)
3. *Calculation of the Proximity index for a comparison of the relevant theories is done with regard to the same stimulus states.* This requirement makes possible a comparison between the relevant theories on the same empirical base and avoids a discussion of the relative importance of the empirical findings themselves. As may be seen from Figure 7.2, the four theories are compared with regard to four stimulus states: S_1, S_2, S_3, S_4.

The proposal for a Proximity index raises some interesting questions, which I shall answer briefly. First, is this index a necessary or sufficient condition for the approximation of a theoretical structure to the URP? It is a necessary but not sufficient condition, because if $P \neq O$, the studied theoretical structure is incorrect, but if $P = O$, it does not mean that the studied theoretical structure is certainly correct. The Proximity index deals with several theoretical structures that are tested by a number of empirical tests. In this sense, the present idea follows Popper (1972), who writes: "And if we fail to refute the new theory, especially in the fields in which its predecessor has been refuted, then we can claim this as one of the objective reasons for *the conjecture that the new theory is a better approximation to the truth than the old theory*" (p. 81, emphasis in the original).

Secondly, does the Proximity index serve scientific progress? The answer is affirmative. If scientific progress takes place when we succeed in proposing theoretical structures that explain a larger number of behavioral phenomena—namely, we succeed in proposing theoretical structures whose Proximity index constantly decreases—we succeed in getting ever closer to the Unknown Real-Process, and the efforts to unveil it seem to be proceeding well.

Thirdly, isn't the Proximity index based on too narrow a view of assessment of a scientific theory? Essentially, the present index rests on empirical success and does not consider other important factors for assessing a theory such as prior theoretical/empirical background, simplicity, and unification (see Zamora Bonilla, 2000; Lipton, 2005; Nola & Sankey, 2007). As for prior theoretical/empirical background, the Proximity index indeed does not in itself take this factor into account, as several other proposals for truthlikeness do. However, while some critiques of these proposals have to do with the great difficulty in defining theoretical/empirical background so as to build on it a truthlikeness approach, the present index may be applied to empirical data directly and without much difficulty. Furthermore, since the Proximity index is anchored to the GEP and this procedure is linked to other research tools such as the H-D method, it may be said that research that tests whether a certain theoretical structure explains the studied phenomenon takes into account the theoretical/empirical background (as stated, the introduction to every research article opens with a review of relevant literature). Regarding simplicity and unification, it may be suggested, on the one hand, that as research on the given phenomenon progresses, so new and improved theoretical structures arise that are also likely to meet these criteria better. On the other hand, if the researcher has to choose between two theories that the Proximity index grades as equal, s/he may choose the simpler one and the one more embedded in the existing theoretical/empirical background. (This notion parallels Lipton [2001a, 2001b, 2005], who in a discussion of IBE suggested that criteria such as simplicity and unification guide the researcher in the choice of the best causal hypothesis for explaining the studied phenomenon.)

THE GENERALITY OF THE GEP

The generality of the GEP is expressed in two ways. First, the concept of theoretical structure that appears on the explanation level can include different hypotheses, theories, and mechanisms (see Figure 7.1); secondly, the GEP can encompass different schemes or models of explanation when they match the Explanatory Characterization (see above). To support this, I examine three research examples whose explanatory analyses accord with the GEP. In the first example (animal learning), I show that three explanation models discussed in the literature (the D-N model, the I-S model, and the S-R model) do not fit the theoretical structure (the "matching law") or the behavioral phenomenon under study. In the second example (face perception and recognition), I show that the present area of research can be handled by the GEP. Finally, in the last example (purposive behavior), I show that the will/belief model of explanation is appropriate for proposing an account for purposive behavior, and that this kind of explanation can be incorporated by the GEP.

Animal Learning

In 1961, Herrnstein reported the results of a learning experiment with hungry pigeons, which led to the discovery of the "matching law." The pigeons learned to peck two keys to get reinforcements: pecking each key was accompanied by a different number of reinforcements. The matching law that Herrnstein discovered is

$$B_r/(B_r + B_l) = R_r/(R_r + R_l)$$

where R_r is the number of reinforcements given for the right-hand key, R_1 is the number of reinforcements given for the left-hand key, B_r is the number of peckings (i.e., behavior) on the right-hand key, and B_1 is the number of peckings on the left-hand key.

Since the publication of that paper, comprehensive empirical and theoretical research has been done on the subject, which has led to extremely interesting developments. For example, it has been found that the matching law did not predict well all the experimental results. Therefore, researchers suggested a relatively new formula, the generalized matching law, which handled the findings better. Moreover, remarkable theoretical efforts have been made to anchor the matching law to wider theoretical frameworks such as economic principles (maximization) (see Baum, 1974; Davison & McCarthy, 1988; McDowell, 2005).

To grasp the connection between the matching law and the GEP, I shall try to answer the question that arises from the following hypothetical experiment. A hungry rat has to choose between two pedals: pressing the right-hand pedal gives the rat eight reinforcements (grains of food) for ten presses, while pressing the left-hand pedal gives the rat four reinforcements for ten presses. The hypothetical result shows that the rat learned to press the right-hand pedal 67 percent of the time and the left-hand pedal 33 percent of the times. How is this behavior to be explained? The answer, namely, the explanation for the rat's behavior, is obtained through the generation of a prediction from the matching law: the percentage of presses on the right-hand pedal is 67 percent because the chance of obtaining reinforcements for pressing this pedal is 8/(8 + 4) = .67. And because there is a match between the prediction and the experiment result, we take the explanation for the rat's behavior by the matching law to be successful. Given this, the following question arises: Which scheme, model of explanation, have we used here? (As stated earlier, the law in itself does not provide an explanation, and an explanation model is required that will use the law to propose an explanation.)

It may be suggested that the explanation for the present experiment in learning follows the D-N model proposed by the philosophers of science Hempel and Oppenheim (1948) (see Hempel, 1965, 1966; Salmon, 1990; chapter 2). Is the D-N model appropriate for explaining the rat's behavior? Beyond the argument that the matching law can be conceived as no more than an empirical generalization that itself requires an explanatory anchorage (see Cummins, 2000), the following doubt arises. Formally, it seems that the explanation by the matching law is the following: one places in the D-N model the matching law, one places in this law the specific facts (the experimental condition 8/[8 + 4] reinforcements), and one obtains the prediction that the rat will press the right-hand pedal 67 percent of the time. Nevertheless, the answer to the above question is negative. As seen, the D-N model offers a prediction of a single event through a deterministic law. By contrast, the matching law concerns allocation of reinforcements and responses between the two pedals as follows: 67 percent of the reinforcements and the responses are linked to the right-hand pedal and 33 percent to the left-hand pedal. That is, the matching law concerns statistical behavior: distribution of the rat's responses between the two pedals, right and left.

Hence, it may be suggested that the explanation model suitable for the matching law is the Inductive-Statistical (I-S) model proposed by Hempel (1965). I shall substantiate this model with an example (see chapter 2). From the statistical law: everyone who smokes two packs of cigarettes a day has a *very high chance* of getting lung cancer, and from the fact that for many years Danny has chain-smoked (four packs a day), a prediction may be suggested that the likelihood of Danny getting cancer is very high. Sadly, Danny did get lung cancer, and the explanation for this rests on this statistical law. However, using the I-S model in the present case too is problematic. Hempel's requirement of this model, which he conceived as based on an inductive argument, is that the statistical law must impart to the explained phenomenon very high inductive probability (it has to be close to 1 for this model to be similar to the D-N model, which generates a prediction with complete certainty).

This requirement is not apt for cases where it is attempted to explain phenomena with low probability of occurrence, for example, in the present case of learning ($p = .67$). To emphasize this point, consider the following fictitious example. It has been found that in a certain city, called city A, the likelihood of road accidents is indeed low (say $p = .200$) but far higher than the national likelihood of road accidents ($p = .002$). How can one explain this phenomenon? Clearly the I-S model is not appropriate for this case. To resolve such problems, Salmon (1971, 1990) proposed the Statistical-Relevance (S-R) model, where the explanation is not based on argument (deductive or inductive) or on high probability of occurrence, but on change in probability of occurrence of the studied phenomenon as a result of discovery of the factors relevant to its happening. For example, in the present road accident instance, it turned out that in city A the traffic lights were broken, so [*p*(accident/broken traffic lights) = .200] > [*p*(accident) = .002]. Given this, one may propose that the S-R model better fits animal learning. Unfortunately, here too a problem arises. It is indeed possible to suggest with this model that the probability of pressing the right-hand pedal is greater than that of pressing the left-hand pedal given the relevant statistical factor (distribution of reinforcements), but it is hard to move from this to an exact statistical prediction (percentage presses on the right-hand pedal is 67 percent) because the S-R model does not refer directly to the matching law itself.

In light of this discussion, the question arises: What, then, is the explanatory procedure suitable for animal learning? To answer, let us look first at the research procedure by means of which the matching law was discovered in Herrnstein's research. This law was discovered when the experiment's results were presented on a set of axes $Y = f(X)$, where there appeared on the X axis %reinforcement on key A (0 percent to 100 percent) and on the Y axis %response on key A (0 percent to 100 percent). A linear function was obtained, which describes the matching law (Herrnstein, 1961, Figure 1). So now, given this description, I propose the following answer: the GEP is the explanatory procedure most suited to dealing with the present finding because it takes into account the matching law and the entire observation range of the studied phenomenon (from 0 percent to 100 percent). We predict rationally the statistical behavior of animals by the matching law inserted into the general concept of theoretical structure. And because the prediction is successful, we suggest that the matching law is an approximation to a URP responsible for the studied behavior. That is, the statistical behavior is determined by distribution of reinforcements, so the matching law allows a satisfactory explanation for the studied behavior over all its probabilistic values from 0 to 1.

Face Perception and Recognition

Research on face perception and recognition concerns four kinds of facial information: featural (eyes, nose, mouth), relational (eyes above nose, nose above mouth), configural (space between eyes, space between nose and mouth), and holistic (perception of the face as a whole unit). One hypothesis that arises from this research, the Configural Processing hypothesis, suggests that all four kinds of information undergo appropriate processing by the cognitive system when the face is presented upright (hair above, chin below). But when the face is presented inverted (chin above, hair below) the processing becomes differential: configural information processing is greatly impaired, but featural information processing is not. (The literature also deals with a similar hypothesis, the Holistic hypothesis, whereby inversion impairs the processing of the whole face. Here, however, I do not treat this hypothesis.) The Configural Processing hypothesis is proposed as the explanation for the Face-Inversion Effect (FIE): perception and recognition of upright faces is much better than it is of inverted faces (e.g., Maurer, Le Grand, & Mondloch, 2002; Rakover, 2002a, 2012a, 2013; Rakover & Cahlon, 2001; Rossion, 2008, 2009). (Actually, FIE also includes the finding that the effect of inversion is greater in faces than in objects and nonfacial shapes. But for our purposes here, it is sufficient to describe the effect of inversion on face perception and recognition.)

In a series of studies, Rakover (2012a) and Rakover and Teucher (1997) showed that the impairment of configural information processing resulting from face inversion is not a necessary condition for the FIE. In these experiments, use was made of the theory-testing method as follows: it was predicted that if the Configural Processing hypothesis is correct, the FIE will not be obtained in isolated features—a prediction that was disconfirmed. These findings and others reviewed in Rakover (2013) undermined the empirical support for the Configural Processing hypothesis. As a result, Rakover (2013) developed the Face-Scheme Incompatibility (FSI) model, which provides explanations for most of the relevant findings associated with FIE, encompasses the relevant theoretical basis, and suggests certain new directions. The FSI model is based on three stages: in stage 1 a face stimulus is formed from the visual stimulus; in stage 2 facial schemes and prototypes are formed from the distributions of face stimuli; and in stage 3 a meaning-bearing face is formed from the interaction between a new incoming face stimulus and the facial schemes and prototypes.

The FSI model is developed within the information-processing theoretical framework, which may be seen as anchored to the general scheme of giving explanations by a description of a cognitive mechanism (e.g., Bechtel & Wright, 2009; Rakover, 2007; Wright & Bechtel, 2007; see chapter 3). In the present case, it is assumed that a cognitive mechanism of visual information processing is constructed of different components that are arranged and act on each other so as to produce FIE. This mechanism represents reality (e.g., pictures of faces and their parts) and processes these representations (e.g., by coding, symbol manipulation, storage, and retrieval of facial information), an operation expressed in the individual's behavior (FIE). The assumption is that the description of the mechanism (components and their action) reflects the activity of the real cognitive system when the individual processes visual information (faces). While the individual is not conscious of the information processing, he is conscious of the studied phenomenon (is aware that a face is being presented to him).

This mechanism handles visual information in a way that attempts to reflect the real activity of the brain/consciousness, and it fits the Explanatory Characterization and can be encompassed by the GEP.

Purposive, Will/Belief, Explanation

David, who lives in Haifa, announced that he wanted to hear the great soprano, the diva, who was due to give a concert at the Opera House in Tel Aviv a month ahead. And indeed, at the due time, David traveled in his car from Haifa to Tel Aviv, went into the Opera House, and enormously enjoyed the diva's singing. How should we explain David's behavior, the ride to Tel Aviv? The answer is provided by applying to this specific behavior the scheme of purposive explanation, [Will/Belief]: if *X wants G* and *believes* that behavior *B* will realize his/her will, then *X* will perform *B* (see chapter 6). David *wished* with all his might to hear the diva's singing, *believed* that the journey to Tel Aviv would fulfill his wish, so he drove to Tel Aviv. We take this specific will/belief explanation as a good one, satisfactory, because there is a match between the prediction and the behavior.

One possible interpretation of [Will/Belief] is to view it as a hypothetical "mental mechanism" in a way similar to the hypothetical information-processing mechanism: it attempts to mirror one's mind activity, and is composed of different components such as will, belief, and rational deliberations, which interact with each other, and finally produce the behavior that realizes one's will.

Given that the individual behavior is predictable, that the individual will perform a certain action due to rational considerations, and that this action will realize his/her will, we may say that this kind of reasoning fits the Explanatory Characterization and can be included in GEP. In other words, if we accept that the way the individual chooses to realize his/her will is rational—namely, he/she chooses the way to realize his/her will through practical rational considerations—we have to accept that his/her behavior can be predicted on rational grounds, and hence accept that the GEP encompasses [Will/Belief].

In sum, these three cases illustrate how well the GEP can be applied to different areas of research in psychology. In the first case, it was shown that although several explanation models did not fit animal learning, the GEP did. In the second case, it was shown that the FSI model, which successfully accounted for the Face-Inversion Effect, can be encompassed by the GEP. In the third case, it was shown that [Will/Belief], as a special purposive explanation scheme, can also be encompassed by the GEP.

DISCUSSION

In this section I discuss four topics of importance for the chapter, touching on broad philosophical ideas: the "Triple Matching Problem"—the problem of matching among *theoretical structures*, *procedures of explanation*, and *the studied phenomena*; the relations between the GEP and other explanatory models; the H-D method and its criticisms; and the realistic approach to GEP.

The Triple Matching Problem

There is a rough relation between theoretical structures and explanatory procedures, which may be illustrated as follows. Theoretical structures including deterministic laws are likely to match a procedure of the D-N model type; statistical laws are likely to match procedures of the I-S model or the S-R model type; causal factors are likely to match procedures of the Causal-Mechanical model type or the use of mechanistic explanations; and a description of cognitive processes is likely to match computational procedures analogous to the operations of a computer. In these cases, the rational relation among the variables of the theoretical structure creates a constraint that dictates the kind of explanatory procedure. But in fact this matching is far more complex, because account has to be taken also of the type of phenomenon under study. So a procedure for explanation must be matched to a theoretical structure and the latter to the type of phenomenon under study. To support this assertion, consider the following two examples.

The first concerns the matching law. As I showed earlier, the properties of animal learning did not accord with the D-N model or even with the I-S model and the S-R model, but did accord with the use of the GEP, because it was possible to set in this procedure (in the theoretical structure) the matching law, which explained the phenomenon of learning across the entire range of its variability.

The second example concerns purposive explanation. Here I shall deal with the following question: Is it possible to explain behavior characterized as purposive by procedures different from [Will/Belief]? My answer is negative (for discussions, see Rakover, 2007, 2011/2012a, 2011/2012b; chapter 6). To further substantiate this answer, I shall consider the following example. In the last decade, there has been a flow of research studies attesting that it is hard to explain behavior without taking into account mental processes having to do with future occurrences, that is, with teleological (purposive) explanations (see Seligman, Railton, Baumeister, & Sripada, 2013). The theoretical concept concerned with this kind of behavior is called "prospection," and is defined by Seligman et al. (2013, p. 120) as "the mental simulation of future possibilities." This explanatory concept refers to a mental process in which the individual tests for a given situation a number of possibilities of future action and chooses on the basis of rational considerations, according to her wishes and values, the best possibility. In essence, then, prospection might be regarded as a process of the kind: if *X*, then *Y* conditional. This explanation is different in several respects from other explanations that appear in psychology: it is characterized as subjective and teleological. Psychology (behaviorism and also cognitive psychology) has rejected explanations of this kind and preferred the sort of explanations accepted in the natural sciences, characterized as objective and causal. Nevertheless, these researchers suggest that prospection may well function as an important explanatory principle that orders and unites different kinds of behavior such as perception, cognition, memory, emotion, motivation, and action.

A problem similar to the Triple Matching Problem was handled by Weber, Van Bouwel, and De Vreese (2013), who developed the "pragmatic approach to scientific explanation." They attempted to coordinate the phenomena to be accounted for, explanation-seeking questions, different formats for explanations, and a set of evaluative questions. To do so, they developed an appropriate explanatory and evaluative toolbox.

The GEP and Other Explanatory Models

Several explanation models suited the Explanatory Characterization of the GEP. Can these models be abandoned, and the GEP be used in their stead? My answer is as follows. These models may be viewed as proposals for different attributes of the Unknown Real-Process and the way it functions. While the GEP is a general procedure, which suggests only that an explanation rests on the use of rational relations, these explanation models describe the rational relations in as much detail as possible: a deductive or inductive argument, a causal process, a statistical process, a mechanism analogous to the operation of the computer, practical considerations for realizing a goal, and so on. The detailing has two implications. On the one hand, the specification allows a better solution to the "Triple Matching Problem," that is, a better match among the following three important elements: the studied phenomenon, the theoretical structure, and the explanatory procedure (the explanation model). On the other hand, the detailing allows the detection of the specific explanation model's limitations (e.g., it does not fit the studied phenomenon). In the case of Herrnstein's matching law, I showed that it was hard to find an explanation model that matched the phenomenon under study (at last this has been achieved by the GEP). Similarly, in the case of purposive behavior, it seems that the best account is provided by employing the will/belief explanation, which can be incorporated by the GEP.

The H-D Method

The chapter has focused on two research tools, two research methods: the GEP and the H-D method. Here I concentrate on the latter, because the GEP is methodologically connected to the H-D method, and because this method has been subjected to serious criticism that must be countered, at least in part (see Glymour, 1980; Haig, 2005; Lipton, 2005; Nola & Sankey, 2007; Rakover, 2003; Sankey, 2008). I shall start by centering on the two most important properties of the H-D method that can assist in its defense against critique.

First, as a procedure for an empirical test, this research tool does not determine the kind and nature of the theoretical structure set in it. In this sense, the theoretical structure can be a conjunction of hypotheses, and at least one of which, so say the critics, is liable to be inappropriate (e.g., H_1 and H_2, where H_1: the configural processing, and H_2: during the day there is no shadow). In practice, however, this is not so. Researchers do not test a hypothesis to which is attached another absurd hypothesis, because they use different research tools to discard conjuncts that lack a theoretical and empirical basis. Given a behavioral phenomenon, the research sets in the H-D method the best theoretical structure, namely, that which meets the theoretical and empirical criteria relevant to the phenomenon under study.

Secondly, practically, when scientists use the H-D method, they use various rational rules to get to the prediction from the theoretical structure, for example, deductive, inductive, mathematical, probabilistic, causal, analogical, and correlational. The answer to the question of how the prediction arises from the theoretical structure is largely determined by the kind of theoretical structure itself (see above). On this matter, Nola and Sankey (2007) write: "The test consequences need not be obtained only by deduction; if the hypotheses are statistical then the consequences are inferred by non-deductive, or inductive, reasoning. So a better name might be *the hypothetico-inferential method*, to cover both cases, deductive and non-deductive inference" (p. 170, emphasis in the original).

To illustrate these two remarks, I turn again to research in face recognition (Rakover, 2012a). The aim of the research was to test the Configural Processing hypothesis described above. The empirical test was based on the following logic:

1. If the Configural Processing hypothesis is correct, the following prediction is correct: an isolated feature presented inverted will not show a FIE-like behavior.
2. The research findings have shown that an FIE-like behavior has been obtained in an isolated feature.

Therefore, the Configural Processing hypothesis is disconfirmed.

The prediction in (1) is based on an inductive inference that inversion of the face impairs processing of configural information (the spatial relations among the facial features). Therefore, when an isolated facial feature is presented inverted, its processing is not expected to be impaired because the configural information has been removed.

As far as I know, no researcher has presented the Configural Processing hypothesis in conjunction with any kind of hypothesis merely because with the H-D method it is possible to use a combination of hypotheses. The research methodology does not permit this, because the researcher must justify the use of the studied hypothesis. Yet note here another related problem.

Every empirical test of a prediction inferred from a research hypothesis requires auxiliary hypotheses by which the experiment can be realized, for example, hypotheses regarding cooperation of the participants, working order of the lab equipment, including the appropriateness of the operational definitions. Thus, the empirical test of the research hypothesis is conducted in conjunction with all the auxiliary hypotheses, a circumstance that creates an interesting problem, which has to do with the case where the prediction has failed. In this case, the research hypothesis and the auxiliary hypotheses together are disconfirmed, and we do not know which is responsible for the failure—the research hypothesis or any of the auxiliary hypotheses. This problem is known as Duhem's problem (see Duhem, 1966; Rakover, 2003).

A practical solution may be offered for this problem by the use of various research tools to ensure that the results of the experiment are relevant to the research hypothesis and not to the auxiliary hypotheses. One may use prior theoretical and empirical knowledge that shows that the auxiliary hypotheses are reliable and valid and, therefore, are not responsible for the failure. An example is the indexes used in testing recognition memory: researchers can safely rely on these indexes since they are based on deep theoretical grounding (e.g., the signal detection theory) and broad empirical support due to a great number of experiments in diverse research fields in psychology. Something similar may be said regarding the experiment's instruments and design, the statistical analysis and additional research tools anchored to prior theoretical knowledge and that can be assessed against accepted standards. In short, the use of these research tools may justify the claim that the experiment's result is relevant to the research hypothesis and not to the auxiliary hypotheses. (Rakover, 2003, developed a pragmatic approach to resolving Duhem's problem, which, however, goes beyond the present discussion.)

As may be seen, the solution to several problems concerning the H-D method is based on the researcher's practical use of the research tools at her disposal. To further substantiate this practical approach, I shall examine the approach of Lipton (2005), who argued in his book *Inference to the Best*

Explanation (IBE) that IBE is better than the H-D method. He also compared the advantages and disadvantages of IBE with the Bayesian approach to confirming hypotheses, and debated this matter with Salmon (see Lipton, 2001a, 2001b, and Salmon, 2001a, 2001b). Although Lipton and Salmon concurred on the critiques of the H-D method, Salmon held that the Bayesian approach surpassed that of the IBE. (The Bayesian approach to confirmation of hypotheses is based on Bayes's famous formula that given a certain hypothesis H and observation O, then $P[H/O] = [P(O/H) \times P(H)]/P[O]$, where the probabilities are interpreted here as degrees of belief. By this approach, O confirms H when the posterior probability of H [$P(H/O)$] is greater than the prior probability of H [$P(H)$].)

In view of the foregoing, I would like to make the following proposition counter Lipton showing that the H-D method has an important and central place in the research process. Lipton (2005) defines IBE thus:

> According to Inference to the Best Explanation, our inferential practices are governed by explanatory considerations. Given our data and our background beliefs, we infer what would, if true, provide the best of the competing explanations we can generate of those data (so long as the best is good enough for us to make any inference at all). Far from explanation only coming on the scene after the inferential work is done, the core idea of Inference to the Best Explanation is that explanatory considerations are a guide to inference. (p. 56)

The IBE raises two interesting problems. The first: Which explanatory procedure should be used when IBE is followed? This is not a simple matter, as Lipton himself admits that if we use the D-N model, IBE becomes the H-D method. As a solution, he suggests using a causal explanation. However, it should be stressed that not all explanations in science are causal. Furthermore, if we suggest the GEP, which also encompasses causal explanation, as a solution it turns out that the choice of the hypothesis that best explains the phenomenon under study depends on the H-D method because, as stated above, the GEP is methodologically connected to that procedure of testing hypotheses.

Secondly, consider the following situation. Let us assume that the problem of the type of explanation that is suited to IBE has been solved; and that in addition, out of three hypotheses suggested for the studied phenomenon, we have chosen H_1 as the best explanation. The research question is: What do we do next? (Otherwise one may suggest that H_1 is a kind of ad hoc hypothesis.) The standard methodological answer is this: test H_1 by making new predictions. And here arises a point that I would like to highlight: these new predictions are not testable empirically by IBE but by the H-D method, simply because the former is not considered a method designed to test empirically new predictions inferred from a given hypothesis. It is an inference method guided by explanatory considerations. An empirical test is based on the component of prediction–observation comparison. This component is the cornerstone in the H-D method. If IBE cannot avoid this component, it approximates the H-D method. If IBE avoids the component, the question arises: Is it a method for testing hypotheses?

GEP and Realism

The GEP is based on the realistic approach. The premise that on observation level (I) there is a URP responsible for the researched behavior is realistic (see Figure 7.1). Earlier I tried to support this in a specific manner. Now I shall try to anchor it to broader arguments holding that scientific theories are directed, and for the most part get close, to describing real processes occurring in the world. Scientific realism is a complex subject with different variants, and subject to manifold debates with those who criticize it. I cannot review this matter here, so I shall concentrate on two important ideas: (1) the proximity of a theory to reality is supported by the theoretical virtues of breadth and depth (see Thagard, 2007); and (2) the Miracle argument, which is considered one of the most intuitive and persuasive arguments for realism (see, e.g., Chakravartty, 2013; Lipton, 2005; Nola & Sankey, 2007; Psillos, 1999; Putnam, 1975a).

First, Thagard (2007) proposed that a theory grows more approximate to reality when it expands its explanatory coherence by means of broadening the range of explained phenomena (explanation of new phenomena), and by deepening, linked to the discovery of a mechanism that underlies the given theory (a mechanism that explains the theory and also the new phenomena). As illustrations, Thagard suggests among others things molecular biochemistry, which supports, deepens, neuroscience and explains how neurons work—also cognitive psychology, which is helped by, enriched by, and receives a depth dimension from research in neural structures and processes.

Second, the Miracle argument suggests that the success of scientific theories (in the sciences) is so great and impressive that if we do not interpret this success as evidence that these theories indeed approximate the real description of the world, we have no other choice but to interpret it as a miracle. This argument is not some kind of deduction. It is an inductive argument, IBE, whose purpose is to convince that circumstances are indeed such that it would be very strange to think that the successful scientific theories do not approximate what takes place in the world. To think otherwise is to interpret this mighty success as a wonder. We choose, therefore, approximation to reality, and not a miracle, as the best explanation for so amazing a scientific success (for different versions of this argument, see Psillos, 1999).

This argument has an interesting implication for the criticism leveled against IBE (Lipton, 2005; Nola & Sankey, 2007; van Fraassen, 1980). Given a number of hypotheses that seem to account for a certain phenomenon, there is no guarantee that among them there is the true hypothesis or the one that approximates reality. That is, it may be suggested that scientists seek the truth under a street lamp, when in fact the truth is located in the dark. Against this, one may say, in keeping with the Miracle argument, that the research process is an ongoing struggle, and in the end when a group of hypotheses does not yield fruit (does not pass the empirical tests), the scientific community begins to turn the spotlight on the darkness until a theoretical structure is found that is sufficiently successful theoretically and empirically to support the belief that this success is actually not a miracle but an approximation to the truth.

How applicable is the Miracle argument to psychology? To my mind, indirectly. This argument concerns theories whose measure of success is enormous, that is, theories that have developed in the sciences. Such a kind of theory does not exist in psychology, which is cut short and fragmental (see Newell, 1973; Rakover, 2012b). The theories in psychology do not carry with them such amazing success. Given this, it would be strange not to state that they are still far away from the truth. But because psychology follows the methodology that prevails in the sciences, a methodology that has eliminated poor theories and has carried on its wings mighty scientific success, it may be suggested that something of this methodological ability to approximate the truth is inspired in research in psychology too.

Finally, since psychology as an empirical science is based on, among other things, analogies (e.g., cognitive psychology is based on the analogy to computer processes), I conclude this chapter with three analogies that may deepen the conviction as to the realistic premise of an Unknown Real-Process. The first is the analogy of the magician: everybody is convinced that there is a logical explanation (often fairly simple) for the stunning performance of the magician, but no one in the crowd can figure out what this logical explanation is. The second is the analogy of code: people in intelligence services have managed to lay hands on an important message from the enemy command, but despite their conviction that the message is coded, they are unable to decipher it. The third analogy is the sculptures of Michelangelo: it is said that the Italian sculptor saw the craft of sculpting as the revelation of the figure hidden in the marble block, a figure cast there by his genius imagination. What have these analogies got to say in respect to scientific research? The scientist presumes that the magic show of nature has an explanation that must be revealed, that this show has a code that must be deciphered, and that these goals will be realized by means of theoretical structures that his imagination casts into the world in which he was born.

References

Baars, B. J. (1988). *A cognitive theory of consciousness.* New York: Cambridge University Press.

Baars, B. J. (2002). The conscious access hypothesis: Origins and recent evidence. *TRENDS in Cognitive Sciences*, 6, 41–52.

Barendregt, M., & van Rappard, H. (2004). Reductionism revisited: On the role of reduction in psychology. *Theory & Psychology,* 14, 453–474.

Bargh, J. A. (2007). Social psychological approaches to consciousness. In P. Zelazo, M. Moscovitch, & E. Thompson (Eds.), *The Cambridge handbook of consciousness.* Cambridge and New York: Cambridge University Press.

Bargh, J. A., & Ferguson, M. J. (2000). Beyond behaviorism: On the automaticity of higher mental processes. *Psychological Bulletin*, 126, 925–945.

Bargh, J. A., & Morsella, E. (2008). The unconscious mind. *Perspectives on Psychological Science*, 3, 73–79.

Baum, W. M. (1974). On two types of deviation from the matching law: Bias and undermatching. *Journal of the Experimental Analysis of Behavior*, 22, 231–242.

Baumeister, R. F. (2008). Free will in scientific psychology. *Perspectives on Psychological Science*, 3, 14–19.

Baumeister, R. F., Vohs, K. D., & Funder, D. C. (2007). Psychology as the science of self-reports and finger movements. *Perspectives on Psychological Science*, 2, 396–403.

Bayne, T. (2009). Consciousness. In J. Symons & P. Calvo (Eds.), *The Routledge companion to philosophy of psychology*. London: Routledge.

Bechtel, W. (2008a). Explanation: Mechanism, modularity, and situated cognition. In P. Robbins & M. Aydede (Eds.), *The Cambridge handbook of situated cognition.* Cambridge, UK: Cambridge University Press.

Bechtel, W. (2008b). Mechanisms in cognitive psychology: What are the options? *Philosophy of Science*, 75, 9830–9994.

Bechtel, W. (2008c). *Mental mechanisms: Philosophical perspectives on cognitive neuroscience*. London: Routledge.

Bechtel, W. (2009). Looking down, around, and up: Mechanistic explanation in psychology. *Philosophical Psychology*, 22, 543–564.

Bechtel, W., & Mundale, J. (1999). Multiple realizability revisited: Linking cognitive and neural states. *Philosophy of Science,* 66, 175–207.

Bechtel, W., & Richardson, R. C. (1993). *Discovering complexity: Decomposition and localization as strategies in scientific research.* Princeton, NJ: Princeton University Press.

Bechtel, W., & Wright, C. D. (2009). What is psychological explanation? In J. Symons & P. Calvo (Eds.), *The Routledge companion to philosophy of psychology.* London: Routledge.

Bell, D. E., Raiffa, H., & Tversky, A. (1988). Descriptive, normative, and prescriptive interactions in decision making. In D. E. Bell, H. Raiffa, & A. Tversky (Eds.), *Decision making: Descriptive, normative, and prescriptive interactions.* Cambridge: Cambridge University Press.

Bem, S., & Looren de Jong, H. L. (1997). *Theoretical issues in psychology: An introduction.* London: Sage.

Bickle, J. (1998). *Psychoneural reduction: The new wave.* Cambridge, MA: The MIT Press.

Bird, A. (2000). *Philosophy of science.* London: Routledge.

Block, N. (1995). The mind as the software of the brain. In E. E. Smith & D. N. Osherson (Eds.), *Thinking: An invitation to cognitive science* (2nd ed., vol. 3). Cambridge, MA: The MIT Press.

Block, N. (1978/1980). Troubles with functionalism. In N. Block (Ed.), *Readings in philosophy of psychology* (vol. 1). Cambridge, MA: Harvard University Press.

Block, N., & Alston, W. P. (1984). Psychology and philosophy. In M. H. Bernstein (Ed.), *Psychology and its allied disciplines.* Hillsdale, NJ: L. Erlbaum.

Bogen, J. (2013). Theory and observation in science. In E. N. Zalta (Ed.), *The Stanford encyclopedia of philosophy* (spring ed.). http://plato.stanford.edu/archives/spr2013/entries/science-theory-observation.

Bokulich, A. (2011). How scientific models can explain. *Synthese*, 180, 33–45.

Bontly, T. (2000). Review of John Bickle's *Psychoneural reduction: The new wave. British Journal for the Philosophy of Science,* 51, 901–905.

Bratman, M. (2001). Taking plans seriously. In E. Millgram (Ed.), *Varieties of practical reasoning.* Cambridge, MA: The MIT Press.

Brook, A., & Stainton, R. J. (2001). *Knowledge and mind: A philosophical introduction.* Cambridge, MA: The MIT Press.

Campbell, N. R. (1953). *What is science?* New York: Dover.

Carroll, J. W. (2010). Laws of nature. In E. N. Zalta (Ed.), *The Stanford encyclopedia of philosophy* (spring edition). http://plato.stanford.edu/archives/spr2010/entries/laws-of-nature.

Causey, R. J. (1972). *Unity of science.* Dordrecht, Holland: D. Reidel.

Chakravartty, A. (2013). Scientific realism. In E. N. Zalta (Ed.), *The Stanford encyclopedia of philosophy* (summer ed.). http://plato.stanford.edu/archives/sum2013/entries/scientific-realism.

Chalmers, D. J. (1996). *The conscious mind: In search of a fundamental theory.* New York: Oxford University Press.

Chalmers, D. J. (1997a). Facing up to the problem of consciousness. In J. Shear (Ed.), *Explaining consciousness—the "hard problem."* Cambridge, MA: The MIT Press.

Chalmers, D. J. (1997b). Moving forward on the problem of consciousness. In J. Shear (Ed.), *Explaining consciousness—the "hard problem."* Cambridge, MA: The MIT Press.

Chomsky, N. (1980). Rule and representation. *The Behavioral and Brain Sciences,* 3, 1–61.

Chow, S. L. (1987). *Experimental psychology: Rationale, procedures and issues.* Calgary, Alberta: Detseling Enterprises.

Churchland, P. M. (1988). *Matter and consciousness* (revised edition). Cambridge, MA: The MIT Press.

Churchland, P. M. (2013). *Matter and consciousness* (3rd ed.). Cambridge, MA: The MIT Press.

Clark, J. M., & Pavio, A. (1989). Observational and theoretical terms in psychology: A cognitive perspective on scientific language. *American Psychologist*, 44, 500–512.

Coombs, C. H., Dawes, R. M., & Tversky, A. (1970). *Mathematical psychology: An elementary introduction*. Englewood Cliffs, NJ: Prentice-Hall.

Copeland, B. J. (1993). *Artificial intelligence.* Oxford: Blackwell.

Copi, I. M. (1968). *Introduction to logic*. New York: Macmillan.

Coren, S., & Girgus, J. S. (1978). *Seeing is deceiving: The psychology of visual illusions.* Hillsdale, NJ: L. Erlbaum.

Cosmelli, D., Lachaux, J-P., & Thompson, E. (2007). Neurodynamical approaches to consciousness. In P. Zelazo, M. Moscovitch, & E. Thompson (Eds.), *The Cambridge handbook of consciousness*. Cambridge and New York: Cambridge University Press.

Crane, T. (1995). *The mechanical mind: A philosophical introduction to minds, machines and mental representation.* London: Penguin Books.

Craver, C. F. (2001). Role functions, mechanisms, and hierarchy. *Philosophy of Science*, 68, 53–74.

Craver, C. F. (2002). Structures of scientific theories. In P. Machamer & M. Silberstien (Eds.), *The Blackwell guide to the philosophy of science.* Malden, MA: Blackwell.

Craver, C., & Tabery, J. (2016). Mechanisms in science. In E. N. Zalta (Ed.), *The Stanford encyclopedia of philosophy* (spring ed.). http://plato.stanford.edu/archives/spr2016/entries/science-mechanisms.

Cummins, R. (1975). Functional analysis. *The Journal of Philosophy*, 72, 741–765.

Cummins, R. (1983). *The nature of psychological explanation.* Cambridge, MA: The MIT Press.

Cummins, R. (2000). "How does it work?" versus "What are the laws?" Two conceptions of psychological explanation. In F. C. Keil & R. A. Wilson (Eds.), *Explanation and cognition.* Cambridge, MA: The MIT Press.

Davidson, D. (1963). Action, reasons and causes. *Journal of Philosophy*, 60, 685–700.

Davidson, D. (1980). *Essays on actions and events.* Oxford: Clarendon Press.

Davison, M., & McCarthy, D. (1988). *The matching law: A research review.* Hillsdale, NJ: Erlbaum.

Dawkins, M. S. (1995). *Unravelling animal behavior* (2nd ed.). Essex: Longman Scientific & Technical.

de Regt, H. W. (2009). Understanding and scientific understanding. In H. W. de Regt, S. Leonelli, & K. Eigner (Eds.), *Scientific understanding: Philosophical perspectives* (pp. 169–186). Pittsburgh: University of Pittsburgh Press.

de Regt, H. W., Leonelli, S., & Eigner, K. (2009a). Focusing on scientific understanding. In H. W. de Regt, S. Leonelli, & K. Eigner (Eds.), *Scientific understanding: Philosophical perspectives* (pp. 1–17). Pittsburgh: University of Pittsburgh Press.

de Regt, H. W., Leonelli, S., & Eigner, K. (2009b). *Scientific understanding: Philosophical perspectives.* Pittsburgh: University of Pittsburgh Press.

Dehaene, S., & Naccache, L. (2001). Towards a cognitive neuroscience of consciousness: Basic evidence and a workspace framework. *Cognition*, 79, 1–37.

Deitel, H. M., & Deitel, B. (1985). *Computers, data and processing.* New York: Academic Press.

Dennett, D. C. (1979). *Brainstorms: Philosophical essays on mind & psychology.* Cambridge, MA: The MIT Press.

Dietrich, E., & Hardcastle, V. G. (2005). *Sisyphus's boulder: Consciousness and the limits of the knowable.* Amsterdam/Philadelphia: John Benjamins.

Dijksterhuis, A., & Aarts, H. (2010). Goals, attention, and (un)consciousness. *Annual Review of Psychology*, 61, 467–490.

Douglas, H. E. (2009). Reintroducing prediction to explanation. *Philosophy of Science*, 76, 444–463.

Dray, W. H. (1957). *Laws and explanation in history.* London: Oxford University Press.

Dray, W. H. (1966). *Philosophical analysis and history.* New York: Harper and Row.

Duhem, P. (1966). *Essays in the history and philosophy of science.* Translated by R. Ariew & P. Baker. Indianapolis and Cambridge: Hackett.

Edelman, D. B., & Seth, A. K. (2009). Animal consciousness: A synthetic approach. *Trends in Neurosciences*, 32, 476–484.

Eigner, K. (2009). Understanding in psychology: Is understanding a surplus? In H. W. de Regt, S. Leonelli, & K. Eigner (Eds.), *Scientific understanding: Philosophical perspectives* (pp. 271–297). Pittsburgh: University of Pittsburgh Press.

Einstein, A., & Infeld, L. (1938/1950). *The evolution of physics.* London: The Scientific Book Club.

Ellis, S. (2008). The varieties of instrumental rationality. *The Southern Journal of Philosophy*, 46, 199–220.

Estes, W. K. (1950). Toward a statistical theory of learning. *Psychological Review*, 57, 94–107.

Evans, J. St. B. T., & Stanovich, K. E. (2013). Dual-process theories of higher cognition: Advancing the debate. *Perspectives on Psychological Science*, 8, 223–241.

Fetzer, J. H. (2001). *Computers and cognition: Why minds are not machines.* Dordrecht: Kluwer Academic.

Feynman, R. P. (1985). *QED: The strange theory of light and matter.* Princeton, NJ: Princeton University Press.

Flanagan, O. (1991). *The science of the mind.* Cambridge, MA: The MIT Press.

Flanagan, O. (1992). *Consciousness reconsidered.* Cambridge, MA: The MIT Press.

Fodor, J. (1974). Special sciences, or the disunity of science as a working hypothesis. *Synthese,* 28, 97–115.

Fodor, J. (1976). *The language of thought.* Sussex: The Harvester Press.

Fodor, J. (1981). *Representations: Philosophical essays on the foundations of cognitive science.* Cambridge, MA: The MIT Press.

Fodor, J. (1983). *The modularity of mind.* Cambridge, MA: The MIT Press.

Fodor, J. (1998). Special sciences: Still autonomous after all these years (a reply to Jaegwon Kim's "M-R and the metaphysics of reduction"). In J. Fodor (Ed.), *In critical condition: Polemical essays on cognitive science and philosophy of mind.* Cambridge, MA: The MIT Press.

Fodor, J., & Pylyshyn, Z. (1988). Connectionism and cognitive architecture: A critical analysis. *Cognition*, 28, 3–71.

Franklin, S. P. (1995). *Artificial minds.* Cambridge, MA: The MIT Press.

Friedman, M. (1974). Explanation and scientific understanding. *Journal of Philosophy*, 71, 5–19.

Funder, D. C. (2009). Naive and obvious questions. *Perspectives on Psychological Science*, 4, 340–344.

Gigerenzer, G., & Goldstein, D. G. (1996). Reasoning the fast and frugal way: Models of bounded rationality. *Psychological Review*, 103, 650–659.

Gijsbers, V. (2007). Why unification is neither necessary nor sufficient for explanation. *Philosophy of Science*, 74, 481–500.

Glymour, C. (1980). *Theory and evidence*. Princeton, NJ: Princeton University Press.

Gollwitzer, P. M. (1999). Implementation intentions: Strong effects of simple plans. *American Psychologist*, 54, 493–503.

Gordon, R. M. (2009). Folk psychology as mental simulation. In E. N. Zalta (Ed.), *The Stanford encyclopedia of philosophy* (fall ed.). http://plato.stanford.edu/archives/fall2009/entries/folkpsych-simulation.

Griffiths, P. E. (1993). Functional analysis and proper functions. *British Journal for the Philosophy of Science*, 44, 409–422.

Guilford, J. P. (1965). *Fundamental statistics in psychology and education* (4th ed.). New York: McGraw-Hill.

Haberlandt, K. (1997). *Cognitive psychology* (2nd ed.). Boston: Allyn and Bacon.

Haig, B. D. (2005). An abductive theory of scientific method. *Psychological Methods*, 10, 371–388.

Hanna, J. F. (1979). An interpretive survey of recent research on scientific explanation. In P. D. Asquith & H. E. Kyburg Jr. (Eds.), *Current research in philosophy of science*. East Lansing, MI: Philosophy of Science Association.

Hanson, N. R. (1958). *Patterns of discovery*. Cambridge: Cambridge University Press.

Harman, G. (1965). The inference to the Best Explanation. *Philosophical Review*, 74, 88–95.

Harnad, S. (1990). The symbol grounding problem. *Physica D,* 42, 335–346.

Hauser, L. (2016). Artificial intelligence. *Internet Encyclopedia of Philosophy,* http://www.iep.utm.edu/art-inte.

Heil, J. (1998). *Philosophy of mind: A contemporary introduction.* London: Routledge.

Heil, J. (2003). Mental causation. In S. P. Stich & T. A. Warfield (Eds.), *The Blackwell guide to philosophy of mind* (pp. 214–234). Malden, MA: Blackwell.

Hempel, C. G. (1942). The function of general laws in history. *Journal of Philosophy*, 39, 35–48.

Hempel, C. G. (1965). *Aspects of scientific explanation and other essays in the philosophy of science.* New York: The Free Press.

Hempel, C. G. (1966). *Philosophy of natural science.* Englewood Cliffs, NJ: Prentice-Hall.

Hempel, C. G. (2001). *The philosophy of Carl G. Hempel: Studies in science, explanation, and rationality.* Edited by J. H. Fetzer. Oxford: Oxford University Press.

Hempel, C. G., & Oppenheim, P. (1948). Studies in the logic of explanation. *Philosophy of Science*, 15, 135–175.

Herrnstein, R. (1961). Relative and absolute strength of response as a function of frequency of reinforcement. *Journal of the Experimental Analysis of Behavior,* 4, 267–272.

Hon, G. (2001). The why and how of explanation: An analytical exposition. In G. Hon & S. S. Rakover (Eds.), *Explanation: Theoretical approaches and applications*. Dordrecht: Kluwer Academic.

Hon, G., & Rakover, S. S. (Eds.) (2001). *Explanation: Theoretical approaches and applications*. Dordrecht: Kluwer Academic.

Horgan, T. (1984). Jackson on physical information and qualia. *The Philosophical Quarterly*, 34, 147–152.

Horgan, T., & Woodward, J. (1985). Folk psychology is here to stay. *The Philosophy Review*, 94, 197–226.

Hull, C. L. (1952). *A behavior system.* New Haven, CT: Yale University Press.

Humphreys, W. C. (1968). Statistical ambiguity and maximal specificity. *Philosophy of Science*, 35, 112–115.

Jackson, F. (1982). Epiphenomenal qualia. *Philosophical Quarterly,* 32, 127–136.

Johnson, R. N. (1972). *Aggression in man and animals*. Philadelphia: W. B. Saunders.

Jones, R. H. (2000). *Reductionism: Analysis and the fullness of reality.* London: Associated University Presses.

Kahneman, D. (2011). *Thinking, fast and slow*. New York: Farrar, Straus and Giroux.

Kaiser, M., & Craver, C. F. (2013). Mechanisms and laws: Clarifying the debate. In H-K Chao, Szu-Ting Chen, & R. L. Millstein (Eds.), *Mechanism and causality in biology and economics* (pp. 125–145). Dordrecht: Springer.

Kanwisher, N., McDermott, J., & Chun, M. M. (1997). The fusiform face area: A module in human extrastraite cortex specialized for face perception. *The Journal of Neuroscience*, 17, 4302–4311.

Keller, L. R. (1989). Decision research with descriptive, normative, and prescriptive purposes—some comments. *Annals of Operations Research*, 19, 485–487.

Khalifa, K. (2013). The role of explanation in understanding. *British Journal for the Philosophy of Science*, 64, 161–187.

Kim, J. (1996). *Philosophy of mind.* Boulder, CO: Westview Press.

Kim, J. (1998). *Mind in a physical world: An essay on the mind-body problem and mental causation*. Cambridge, MA: The MIT Press.

Kim, J. (2002). Précis of mind in a physical world. *Philosophy and Phenomenological Research*, 65, 640–643.

Kitcher, P. (1981). Explanatory unification. *Philosophy of Science*, 48, 507–531.

Kitcher, P. (1989). Explanatory unification and the causal structure of the world. In P. Kitcher & W. Salmon (Eds.), *Scientific explanation* (pp. 410–505). Minneapolis: University of Minnesota Press.

Knobe, J., & Nichols, S. (2008). *Experimental philosophy*. Oxford: Oxford University Press.

Knobe, J., Buckwalter, W., Nichols, S., Robbins, P., Sarkissian, H., & Sommers, T. (2012). Experimental philosophy. *Annual Review of Psychology*, 63, 81–99.

Kouider, S. (2009). Neurobiological theories of consciousness. In W. Banks (Ed.), *Encyclopedia of consciousness*. Oxford: Elsevier Press.

Kuhn, T. S. (1970). *The structure of scientific revolutions* (2nd ed.). Chicago: University of Chicago Press.

Lambert, K., & Brittan, G. G., Jr. (1992). *An introduction to the philosophy of science* (4th ed.). Atascadero, CA: Ridgeview.

Leahey, T. H. (2004). *A history of psychology* (6th ed.). Upper Saddle River, NJ: Pearson/Prentice-Hall.

LeDoux, J. (1996). *The emotional brain: The mysterious underpinnings of emotional life*. New York: Simon & Schuster.

Lenhard, J. (2009). The great deluge: Simulation modeling and scientific understanding. In H. W. de Regt, S. Leonelli, & K. Eigner (Eds.), *Scientific understanding: Philosophical perspectives* (pp. 169–186). Pittsburgh: University of Pittsburgh Press.

Lenman, J. (2011). Reasons for action: Justification vs. explanation. In E. N. Zalta (Ed.), *The Stanford encyclopedia of philosophy* (winter ed.). http://plato.stanford.edu/archives/win2011/entries/reasons-just-vs-expl.

Leuridan, B. (2010). Can mechanisms really replace laws of nature? *Philosophy of Science*, 77, 317–340.

Levine, J. (1983). Materialism and qualia: The explanatory gap. *Pacific Philosophical Quarterly*, 64, 354–361.

Levy, A. (2010). Makes a difference: Review of Michael Strevens' *Depth: An account of scientific explanation*. Cambridge, MA: Harvard University Press. *Biology and Philosophy*, 26, 459–467.

Libet, B. (1985). Unconscious cerebral initiative and the role of conscious will in voluntary action. *The Behavioral and Brain Sciences*, 8, 526–566.

Libet, B., Gleason, C. A., Wright, E. W., & Pearl, D. K. (1983). Time of conscious intention to act in relation to onset of cerebral activity (readiness-potential): The unconscious intention of a freely voluntary act. *Brain*, 106, 623–642.

Lipton, P. (1992). The seductive-nomological model. *Studies in History and Philosophy of Science*, 23, 691–698.

Lipton, P. (2001a). Is explanation a guide to inference? A reply to Wesley C. Salmon. In G. Hon & S. S. Rakover (Eds.), *Explanation: Theoretical approaches and applications*. Dordrecht: Kluwer Academic.

Lipton, P. (2001b). What good is an explanation? In G. Hon & S. S. Rakover (Eds.), *Explanation: Theoretical approaches and applications*. Dordrecht: Kluwer Academic.

Lipton, P. (2005). *Inference to the best explanation* (2nd ed.). London: Routledge.

Lipton, P. (2009). Understanding without explanation. In H. W. de Regt, S. Leonelli, & K. Eigner (Eds.), *Scientific understanding: Philosophical perspectives* (pp. 43–63). Pittsburgh: University of Pittsburgh Press.

Livio, M. (2009). *Is God a mathematician?* New York: Simon and Schuster.

Locke, E. A., & Latham, G. P. (2002). Building a practically useful theory of goal setting and task motivation. *American Psychologist*, 57, 705–717.

Logan, G. D. (1988). Toward an instance theory of automatization. *Psychological Review*, 95, 492–527.

Looren de Jong, H. (2003). Causal and functional explanations. *Theory & Psychology,* 13, 291–317.

Ludwig, K. (2003). The mind-body problem: An overview. In S. P. Stich & T. A. Warfield (Eds.), *The Blackwell guide to philosophy of mind* (pp. 1–46). Malden, MA: Blackwell.

Lycan, W. G. (2003). The mind-body problem. In S. P. Stich & T. A. Warfield (Eds.), *The Blackwell guide to philosophy of mind* (pp. 47–64). Malden, MA: Blackwell.

MacCorquodale, K., & Meehl, P. E. (1948). On a distinction between hypothetical constructs and intervening variables. *Psychological Review*, 55, 95–107.

Machamer, P., Darden, L., & Craver, C. (2000). Thinking about mechanisms. *Philosophy of Science*, 67, 1–25.

Mackie, J. L. (1974). *The cement of the universe: A study of causation*. Oxford: Clarendon Press.

MacMillan, N. A., & Creelman, C. D. (2005). *Detection theory: A user's guide* (2nd ed.). New York: Psychology Press.

Mahajan, G. (2011). *Explanation and understanding in the human sciences* (3rd ed.). Oxford: Oxford University Press.

Marcus, G. F. (2001). *The algebraic mind: Integrating connectionism and cognitive science.* Cambridge, MA: The MIT Press.

Masserman, J. H. (1946). *Principles of dynamic psychiatry*. Philadelphia: Saunders.

Maurer, D., Le Grand, R., & Mondloch, C. J. (2002). The many faces of configural processing. *TRENDS in Cognitive Sciences*, 6, 255–260.

Maxwell, N. (2000). The mind-body problem and explanatory dualism. *Philosophy,* 75, 49–71.

McCauley, R. N., & Bechtel, W. (2001). Explanatory pluralism and heuristic identity theory. *Theory & Psychology*, 11, 736–760.

McDermott, D. (2007). Artificial intelligence and consciousness. In P. Zelazo, M. Moscovitch, & E. Thompson (Eds.), *The Cambridge handbook of consciousness*. Cambridge and New York: Cambridge University Press.

McDowell, J. J. (2005). On the classic and modern theories of matching. *Journal of the Experimental Analysis of Behavior*, 84, 11–127.

McGinn, C. (1989). Can we solve the mind-body problem? *Mind*, 98, 349–366.

McGovern, K., & Baars, B. J. (2007). Cognitive theories of consciousness. In P. Zelazo, M. Moscovitch, & E. Thompson (Eds.), *The Cambridge handbook of consciousness*. Cambridge and New York: Cambridge University Press.

McKone, E., & Robbins, R. (2011). Are faces special? In A. J. Calder, G. Rhodes, M. H. Johnson, & J. V. Haxby (Eds.), *The Oxford handbook of face perception*. Oxford: Oxford University Press.

Michell, J. (1990). *An introduction to the logic of psychological measurement.* Hillsdale, NJ: L. Erlbaum.

Michell, J. (1999). *Measurement in psychology: Critical history of a methodological concept.* Cambridge: Cambridge University Press.

Miller, G. A. (2011). Mistreating psychology in the decades of the brain. *Perspectives on Psychological Science*, 5, 716–743.

Millgram, E. (2001). Practical reasoning: The current state of play. In E. Millgram (Ed.), *Varieties of practical reasoning*. Cambridge, MA: The MIT Press.

Millikan, R. R. (1989). An ambiguity in the notion of function. *Biology and Philosophy*, 4, 172–176.

Mitchell, C. J., De Houwer, J., & Lovibond, P. F. (2009). The propositional nature of human associative learning. *Behavioral and Brain Sciences*, 32, 183–246.

Morrison, M. (2009). Understanding in physics and biology: From abstract to the concrete. In H. W. de Regt, S. Leonelli, & K. Eigner (Eds.), *Scientific understanding: Philosophical perspectives* (pp. 123–145). Pittsburgh: University of Pittsburgh Press.

Morsella, E. (2005). The function of phenomenal states: Supramodular interaction theory. *Psychological Review*, 112, 1000–1021.

Nagel, E. (1961). *The structure of science: Problems in the logic of explanation.* London: Routledge & Kegan Paul.

Naylor, B. G., & Handford, P. (1985). In defence of Darwin's theory. *BioScience*, 35, 478–484.

Neale, J. M., & Liebert, R. M. (1986). *Science and behavior: An introduction to methods of research* (3rd ed.). Englewood Cliffs, NJ: Prentice-Hall.

Newell, A. (1973). You can't play 20 questions with nature and win: Perceptive comments on the papers of this symposium. In W. C. Chase (Ed.), *Visual information processing*. New York: Academic Press.

Newell, A. (1982). The knowledge level. *AI Magazine*, 2, 1–33.

Ney, A. (2016). Reductionism. *Internet Encyclopedia of Philosophy.* http://www.iep.utm.edu/red-ism.

Nisbett, R. E., & Wilson, T. D. (1977). Telling more than we can know: Verbal reports on mental processes. *Psychological Review,* 84, 231–254.

Nola, R., & Sankey, H. (2007). *Theories of scientific method: An introduction.* Stocksfield: Acumen Publishing.

Oddie, G. (2007). Truthlikeness. In E. N. Zalta (Ed.), *The Stanford encyclopedia of philosophy*. http://plato.stanford.edu/entries/truthlikness.

Owens, D. (1989). Levels of explanation. *Mind*, 98, 59–79.

Pachella, R. G. (1974). The interpretation of reaction time in information-processing research. In B. H. Kantowitz (Ed.), *Human information processing: Tutorials in performance and cognition.* New York: L. Erlbaum.

Palmer, S. E. (1999). *Vision science: Photons to Phenomenology*. Cambridge, MA: The MIT Press.

Palmer, S. E., & Kimchi, R. (1986). The information processing approach to cognition. In T. J. Knapp & L. C. Robertson (Eds.), *Approaches to cognition: Contrasts and controversies*. Hillsdale, NJ: L. Erlbaum.

Palmeri, T. J. (2002). Automaticity. In L. Nadel et al. (Eds.), *Encyclopedia of cognitive science* (pp. 390–401). London: Nature.

Papineau, D. (1999). Methodology: The elements of the philosophy of science. In A. C. Grayling (Ed.), *Philosophy 1: A guide through the subject*. Oxford: Oxford University Press.

Penrose, R. (1994). *Shadows of the mind: A search for the missing science of consciousness*. Oxford: Oxford University Press.

Piccinini, G., & Craver, C. (2011). Integrating psychology and neuroscience: Functional analyses as mechanism sketches. *Synthese*, 183, 283–311.

Polger, T. W. (2004). *Natural minds*. Cambridge, MA: The MIT Press.

Popper, K. R. (1963). *Conjectures and refutations*. London: Routledge.

Popper, K. R. (1972). *Objective knowledge: An evolutionary approach*. Oxford: Oxford University Press

Popper, K. R. (1994). *The myth of the framework: In defense of science and rationality*. Edited by M. A. Notturno. London and New York: Routledge.

Popper, K. R. (1995). *The logic of scientific discovery*. London and New York: Routledge.

Potochnik, A. (2011). Explanation and understanding: An alternative to Strevens' *Depth*. *The European Journal for the Philosophy of Science*, 1, 29–38.

Psillos, S. (1999). *Scientific realism: How science tracks truth*. London and New York: Routledge.

Psillos, S. (2002). *Causation and explanation*. Chesham, UK: Acumen.

Psillos, S. (2007). Past and contemporary perspectives on explanation. In T. Kuipers, D. M. Gabbay, P. Thagard, & J. Woods (Eds.), *Handbook of the philosophy of Science: General philosophy of science—focal issues* (pp. 97–173). Amsterdam: Elsevier.

Putnam, H. (1967). Psychological predicates. In W. Capitan & D. Merrill (Eds.), *Art, mind, and religion*. Pittsburgh, PA: University of Pittsburgh Press.

Putnam, H. (1975a). *Mathematics, matter and method*. Cambridge: Cambridge University Press.

Putnam, H. (1975b). *Mind, language and reality: Philosophical papers, V. 2*. Cambridge: Cambridge University Press.

Pylyshyn, Z. W. (1980). Computation and cognition. *The Behavioral and Brain Sciences*, 3, 111–169.

Pylyshyn, Z. W. (1984). *Computation and cognition*. Cambridge, MA: The MIT Press.

Rakover, S. S. (1975). Tolerance of pain as a measure of fear. *Learning and Motivation*, 6, 43–61.

Rakover, S. S. (1983). Hypothesizing from introspections: A model for the role of mental entities in psychological explanation. *Journal for the Theory of Social Behavior*, 13, 211–230.

Rakover, S. S. (1990). *Metapsychology: Missing links in behavior, mind and science*. New York: Paragon/Solomon.

Rakover, S. S. (1993). Empirical criteria for task susceptibility to introspective awareness and awareness effects. *Philosophical Psychology,* 6, 451–467.

Rakover, S. S. (1996). The place of consciousness in the information processing approach: The mental-pool thought experiment. *Behavioral and Brain Sciences,* 19, 535–536.

Rakover, S. S. (1997). Can psychology provide a coherent account of human behavior? A proposed multiexplanation-model theory. *Behavior and Philosophy*, 25, 43–76.

Rakover, S. S. (1999). The computer that simulated John Searle in the Chinese room. *New Ideas in Psychology*, 17, 55–66.

Rakover, S. S. (2002a). Featural vs. configurational information in faces: A conceptual and empirical analysis. *British Journal of Psychology*, 93, 1–30.

Rakover, S. S. (2002b). Scientific rules of the game and the mind/body: A critique based on the theory of measurement. *Journal of Consciousness Studies*, 9, 52–58.

Rakover, S. S. (2003). Experimental psychology and Duhem's problem. *Journal for the Theory of Social Behaviour*, 33, 45–66.

Rakover, S. S. (2005). Two methodologies of memory research: "Explanation-testing" and "reconstruction." In J. G. Clement & M. K. Marks (Eds.), *Computer-graphic facial reconstruction* (pp. 255–283). Amsterdam: Elsevier Academic Press.

Rakover, S. S. (2007). *To understand a cat: Methodology and philosophy*. Amsterdam/Philadelphia: John Benjamins.

Rakover, S. S. (2011/2012a). Methodological dualism and multi-explanation framework: Replies to criticisms and further developments. *Behavior and Philosophy*, 39/40, 107–125.

Rakover, S. S. (2011/2012b). A plea for methodological dualism and multi-explanation framework in psychology. *Behavior and Philosophy*, 39/40, 17–43.

Rakover, S. S. (2012a). A feature-inversion effect: Can an isolated feature show behavior like the face-inversion effect? *Psychonomic Bulletin & Review*, 19, 617–624.

Rakover, S. S. (2012b). Psychology as an associational science: A methodological viewpoint. *Open Journal of Philosophy*, 2, 143–152.

Rakover, S. S. (2013). Explaining the face-inversion effect: The face-scheme incompatibility (FSI) model. *Psychonomic Bulletin & Review*, 20, 665–692.

Rakover, S. S., & Cahlon, B. (1989). To catch a thief with a recognition model: The model and some empirical results. *Cognitive Psychology*, 21, 423–468.

Rakover, S. S., & Cahlon, B. (1999). The Catch Model: A solution to the problem of saliency in facial features. *Spatial Vision*, 12, 73–81.

Rakover, S. S., & Cahlon, B. (2001). *Face recognition: Cognitive and computational processes*. Amsterdam/Philadelphia: John Benjamins.

Rakover, S. S., & Cahlon, B. (2008). A face recognition by similarity (FRBS) conjecture. *Perception & Psychophysics*, 70, 969–982.

Rakover, S. S., & Cahlon, B. (2013). Cognitive processing of scrambled faces. *American Journal of Psychology*, 126, 235–244.

Rakover, S. S., & Teucher, B. (1997). Facial inversion effects: Part and whole relationship. *Perception & Psychophysics*, 59, 752–761.

Reese, H. W. (1989). Rules and rule-governance: Cognitive and behavioristic views. In S. C. Hayes (Ed.), *Rule-governed behavior: Cognition, contingencies, and instructional control*. New York: Plenum Press.

Robb, D., & Heil, J. (2014). Mental Causation. In E. N. Zalta (Ed.), *The Stanford encyclopedia of philosophy* (spring ed.). http://plato.stanford.edu/archives/spr2014/entries/mental-causation.

Rosenberg, A. (1988). *Philosophy of social science*. Boulder, CO: Westview Press.

Roser, M., & Gazzaniga, M. S. (2004). Automatic brains—interpretive minds. *Current Directions in Psychological Science*, 13, 56–59.

Rossion, B. (2008). Picture-plane inversion leads to qualitative changes of face perception. *Acta Psychologica*, 128, 274–289.

Rossion, B. (2009). Distinguishing the cause and consequence of face inversion: The perceptual field hypothesis. *Acta Psychologica*, 132, 300–312.

Saatsi, J., & Pexton, M. (2013). Reassessing Woodward's account of explanation: Regularities, counterfactuals, and noncausal explanations. *Philosophy of Science*, 80, 613–624.

Salmon, W. C. (1971). Statistical explanation. In W. Salmon et al. (Eds.), *Statistical explanation and statistical relevance* (pp. 29–87). Pittsburgh: University of Pittsburgh Press.

Salmon, W. C. (1978). Why ask "why?"? An inquiry concerning scientific explanation. *Proceedings and Addresses of the American Philosophical Association*, 51, 683–705.

Salmon, W. C. (1984). *Scientific explanation and the causal structure of the world*. Princeton: Princeton University Press.

Salmon, W. C. (1990). *Four decades of scientific explanation*. Minneapolis: University of Minnesota Press.

Salmon, W. C. (1997). Causality and explanation: A reply to two critics. *Philosophy of Science*, 64, 461–477.

Salmon, W. C. (2001a). Explanation and confirmation: A Bayesian critique of Inference to the Best Explanation. In G. Hon & S. S. Rakover (Eds.), *Explanation: Theoretical approaches and applications*. Dordrecht: Kluwer Academic Publishers.

Salmon, W. C. (2001b). Reflections of a bashful Bayesian: A reply to Peter Lipton. In G. Hon & S. S. Rakover (Eds.), *Explanation: Theoretical approaches and applications*. Dordrecht: Kluwer Academic Publishers.

Salmon, W. C., Jeffrey, R. C., & Greeno, J. G. (1971). *Statistical explanation and statistical relevance*. Pittsburgh: University of Pittsburgh Press.

Samuels, R., Stich, S., & Faucher, L. (2004). Reason and rationality. In I. Niiniluoto, M. Sintonen, & J. Wolenski (Eds.), *Handbook of epistemology*. Dordrecht: Kluwer.

Sankey, H. (2008). Scientific method. In S. Psillos & M. Curd (Eds.), *Routledge companion to philosophy of science*. Abingdon: Taylor & Francis.

Saygin, A. P., Cicekli, I., & Akman, V. (2000). Turing test: 50 years later. *Minds and Machines*, 10, 463–518.

Sayre-McCord, G. (1989). Functional explanations and reasons as causes. In J. E. Tomberlin (Ed.), *Philosophical Perspectives. 3: Philosophy of Mind and Action Theory*. Atascadero, CA: Ridgeview.

Schaffner, K. F. (1967). Approaches to reduction. *Philosophy of Science*, 34, 137–147.

Schaffner, K. F. (1993). *Discovery and explanation in biology and medicine.* Chicago: University of Chicago Press.

Schneider, W. (2009). Automaticity and consciousness. In W. Banks (Ed.), *Encyclopedia of consciousness*. Oxford: Elsevier Press.

Scholl, B. J., & Tremoulet, P. D. (2000). Perceptual causality and animacy. *TRENDS in Cognitive Sciences*, 4, 299–309.

Schueler, G. F. (2003). *Reasons and purposes: Human rationality and the teleological explanation of action*. Oxford: Oxford University Press.

Schurz, G., & Lambert, K. (1994). Outline of a theory of scientific understanding. *Synthese*, 101, 65–120.

Schweigert, W. A. (2006). *Research methods in psychology: A handbook.* Long Grove, IL: Waveland Press.

Scriven, M. (1959). Explanation and prediction in evolutionary theory. *Science*, 130, 477–482.

Scrödinger, E. (1992). *What is life? The physical aspects of the living cell.* Cambridge: Cambridge University Press.

Searle, J. R. (1980). Minds, brains and programs. *The Behavioral and Brain Sciences*, 3, 417–457.

Seligman, M. E. P., Railton, P., Baumeister, R. F., & Sripada, C. (2013). Navigating into the future or driven by the past. *Perspectives on Psychological Science*, 8, 119–141.

Shafir, E., & LeBoeuf, R. A. (2002). Rationality. *Annual Review of Psychology*, 53, 491–517.

Shear, J. (Ed.) (1997). *Explaining consciousness—the "hard problem."* Cambridge, MA: The MIT Press.

Silberstein, M. (2002). Reduction, emergence and explanation. In P. Machamer & M. Silberstien (Eds.), *The Blackwell Guide to the Philosophy of Science.* Malden, MA: Blackwell.

Simard, T. G., & Basmajian, J. V. (1967). Methods in training the conscious control of motor units. *Archives of Physical Medicine Rehabilitation*, 48, 12–19.

Simon, H. A. (1991). Cognitive architectures and rational analysis: Comment. In K. VanLehn (Ed.), *Architectures for intelligence*. Hillsdale, NJ: L. Erlbaum.

Simon, H. A. (1992). What is an "explanation" of behavior? *Psychological Science*, 3, 150–161.

Sklar, L. (1967). Types of inter-theoretic reductions. *British Journal for the Philosophy of Science,* 18, 109–124.

Skow, B. (2014). Are there non-causal explanations (of particular events)? *The British Journal for the Philosophy of Science,* 65, 445–467.

Skow, B. (2015). Scientific explanation. In P. Humphreys (Ed.), *The Oxford handbook of the philosophy of science*. Oxford: Oxford University Press.

Smolensky, P. (1988). On the proper treatment of connectionism. *Behavioral and Brain Sciences,* 11, 1–74.

Spence, K. W. (1956). *Behavior theory and conditioning.* New Haven, CT: Yale University Press.

Sperling, G. (1960). The information available in brief visual presentations. *Psychological Monographs*, 74(11).

Sternberg, R. J. (2009). *Cognitive psychology* (5th ed.). Belmont, CA: Wadsworth.

Stich, S. P. (1983). *From folk psychology to cognitive science.* Cambridge, MA: The MIT press.

Strevens, M. (2004). The causal and unification approaches to explanation unified—causally. *Nous*, 38, 154–176.

Strevens, M. (2008a). Comments on Woodward, making things happen. *Philosophy and Phenomenological Research*, 75, 171–192.

Strevens, M. (2008b). *Depth: An account of scientific explanation.* Cambridge, MA: Harvard University Press.

Strevens, M. (2013). No understanding without explanation. *Studies in History and Philosophy of Science*, 44, 510–515.

Sturm, T. (2012). The "rationality wars" in psychology: Where they are and where they could go. *Inquiry*, 55, 68–81.

Sun, R., & Franklin, S. (2007). Computational models of consciousness: A taxonomy and some examples. In P. Zelazo, M. Moscovitch, & E. Thompson (Eds.), *The Cambridge handbook of consciousness.* Cambridge and New York: Cambridge University Press.

Suppe, F. (1977a). Introduction. In F. Suppe (Ed.), *The structure of scientific theories.* Urbana: University of Illinois Press.

Suppe, F. (1977b). *The structure of scientific theories.* Urbana: University of Illinois Press.

Swartz, N. (1985). *The concept of physical law.* Cambridge: Cambridge University Press.

Thagard, P. (1996). *Mind: Introduction to cognitive science.* Cambridge, MA: The MIT press.

Thagard, P. (2007). Coherence, truth, and the development of scientific knowledge. *Philosophy of Science*, 74, 28–47.

Thagard, P. (2008). Cognitive science. In S. Psillos & M. Curd (Eds.), *Routledge companion to philosophy of science.* Abingdon: Taylor & Francis.

Thagard, P., in collaboration with others (2012). *The cognitive science of science: Explanation, discovery, and conceptual change.* Cambridge, MA: The MIT Press.

Tolman, E. C. (1959). Principles of purposive behavior. In S. Koch (Ed.), *Psychology: A study of a science* (vol. 2). New York: Appleton-Century-Crofts.

Toulmin, S. (1974). Rules and their relevance for understanding human behavior. In T. Mischel (Ed.), *Understanding other persons.* Oxford: Blackwell.

Trout, J. D. (2002). Scientific explanation and the sense of understanding. *Philosophy of Science*, 69, 212–233.

Turing, A. M. (1950). Computing machinery and intelligence. *Mind*, 59, 433–460.

Tversky, A., & Kahneman, D. (1974). Judgment under uncertainty: Heuristics and biases. *Science*, 185, 1124–1131.

Tye, M. (2015). Qualia. In E. N. Zalta (Ed.), *The Stanford encyclopedia of philosophy.* http://plato.stanford.edu/archives/fall2015/entries/qualia.

Umiltà, C. (2007). Consciousness and control of action. In P. Zelazo, M. Moscovitch, & E. Thompson (Eds.), *The Cambridge handbook of consciousness*. Cambridge and New York: Cambridge University Press.

Uttal, W. R. (2001). *The new phrenology: The limits of localizing cognitive processes in the brain*. Cambridge, MA: The MIT Press.

Uttal, W. R. (2014). Are neuroreductionist explanations of cognition possible? *Behavior and Philosophy*, 41, 37–64.

van Fraassen, B. C. (1980). *The scientific image*. Oxford: Clarendon Press.

Van Gulick, R. (1995). What would count as explaining consciousness? In T. Metzinger (Ed.), *Conscious-experience. Schöningh*: Imprint Academic; Paderborn: Thorverton.

van Riel, R., & Van Gulick, R. (2016). Scientific reduction. In E. N. Zalta (Ed.), *The Stanford encyclopedia of philosophy* (spring ed.). http://plato.stanford.edu/archives/spr2016/entries/scientific-reduction.

Velmans, M. (1991). Is human information processing conscious? *Behavioral and Brain Sciences*, 14, 651–669.

Von Eckardt, B. (1993). *What is cognitive science?* Cambridge, MA: The MIT Press.

von Wright, G. H. (1971). *Explanation and understanding*. London: Routledge and Kegan Paul.

Weber, E., Van Bouwel, J., & De Vreese, L. (2013). *Scientific explanation*. Dordrecht: Springer.

Weidemann, G., Satkunarajah, M., & Lovibond, P. F. (2016). I think, therefore eyeblink: The importance of contingency awareness in conditioning. *Psychological Science*, 27, 467–475.

Weinert, F. (1995). Laws of nature—laws of science. In F. Weinert (Ed.), *Laws of nature: Essays on the philosophical, scientific and historical dimensions*. Berlin and New York: de Gruyer.

Wenger, D. M. (2003). The mind's best trick: How we experience conscious will. *TRENDS in Cognitive Sciences*, 7, 65–69.

Winkielman, P., & Berridge, K. C. (2004). Unconscious emotion. *Current Directions in Psychological Science*, 13, 120–123.

Woodward, J. (2000). Explanation and invariance in the special sciences. *British Journal for the Philosophy of Science*, 51, 197–254.

Woodward, J. (2002). Explanation. In P. Machamer & M. Silberstein (Eds.), *The Blackwell guide to the philosophy of science*. Oxford: Blackwell.

Woodward, J. (2003). *Making things happen: A theory of causal explanation*. New York: Oxford University Press.

Woodward, J. (2008). Response to Strevens. *Philosophy and Phenomenological Research*, 75, 193–212.

Woodward, J. (2011). Scientific explanation. In E. N. Zalta (Ed.), *The Stanford encyclopedia of philosophy* (winter ed.). http://plato.stanford.edu/archives/win2011/entries/scientific-explanation.

Wright, C., & Bechtel, W. (2007). Mechanisms and psychological explanation. In P. Thagard (Ed.), *Philosophy of Psychology and Cognitive Science* (vol. 4). New York: Elsevier.

Yoo, J. (2015). Mental causation. *The Internet Encyclopedia of Philosophy*, ISSN 2161-0002. http://www.iep.utm.edu/,02/11/2015.

Zamora Bonilla, J. P. (2000). Truthlikeness, rationality and scientific method. *Synthese*, 122, 321–335.

Author Index

Subject Index

About the Author

Sam S. Rakover is a professor in the department of psychology at Haifa University, Israel. He has published five books and approximately one hundred papers on animal learning, face recognition, and philosophy of science and mind. He is also the author of ten novels and dozens of short stories (in Hebrew).

www.ingramcontent.com/pod-product-compliance
Lightning Source LLC
LaVergne TN
LVHW020325220926
848484LV00001BB/133
* 9 7 8 1 4 9 8 5 3 6 6 8 4 *